SpringerBriefs in Applied Sciences and Technology

# Continuum Mechanics

W0079086

**Series Editors**

Holm Altenbach, Institut für Mechanik, Lehrstuhl für Technische Mechanik, Otto von Guericke University Magdeburg, Magdeburg, Germany

Andreas Öchsner, Faculty of Mechanical Engineering, Esslingen University of Applied Sciences, Esslingen am Neckar, Germany

These SpringerBriefs publish concise summaries of cutting-edge research and practical applications on any subject of Continuum Mechanics and Generalized Continua, including the theory of elasticity, heat conduction, thermodynamics, electromagnetic continua, as well as applied mathematics.

**SpringerBriefs in Continuum Mechanics** are devoted to the publication of fundamentals and applications, presenting concise summaries of cutting-edge research and practical applications across a wide spectrum of fields. Featuring compact volumes of 50 to 125 pages, the series covers a range of content from professional to academic.

Andreas Öchsner

# The Fundamental Equations of Beams and Plates

## A Comparative Representation

 Springer

Andreas Öchsner
Faculty of Mechanical and Systems
Engineering
Esslingen University of Applied Sciences
Esslingen, Baden-Württemberg, Germany

ISSN 2191-530X          ISSN 2191-5318  (electronic)
SpringerBriefs in Applied Sciences and Technology
ISSN 2625-1329          ISSN 2625-1337  (electronic)
SpringerBriefs in Continuum Mechanics
ISBN 978-3-031-76275-8          ISBN 978-3-031-76276-5  (eBook)
https://doi.org/10.1007/978-3-031-76276-5

This Springer imprint is published by the registered company Springer Nature Switzerland AG
The registered company address is: Gewerbestrasse 11, 6330 Cham, Switzerland

If disposing of this product, please recycle the paper.

# Preface

Beam and plate elements constitute important structural members, which can be found in quite different application areas, ranging from automotive, to aerospace or civil engineering structures to name only a few. The classical bachelor's degree programs in mechanical engineering cover at least the so-called thin beam during the first year of education in applied mechanics. If the thick version of a beam or the two-dimensional generalizations, i.e., thin and thick plates, are covered, depends on the program and the corresponding specializations. Nevertheless, these one-dimensional, i.e., thin and thick beams, and two-dimensional members, i.e., thin and thick plates, are very important for many technical applications. Thus, mechanical and civil engineers, also at the beginning of their career, must be able to handle and apply these structural members.

In this context, this concise teaching book has its focus with the following intentions: To provide a comparative description of the fundamental equations of beams and plates, which allows a relatively simple transfer of the knowledge from beams to plates and/or from thin to thick members. This should make it possible to quickly acquire and understand the mechanical description of structural members, which are not covered in a degree program. For those who already know these structural members, this teaching book provides a uniform systematic description based on differential operator symbols to illustrate the similarities and differences in the description of the fundamental equations. Thus, it may offer a new, i.e., more formal, view on the modeling of these structural members.

Concerning the mechanical modeling of the mentioned structural members, a classical approach is followed where the three basic equations of continuum mechanics, i.e., the equilibrium equation, the kinematics relation and the constitution are first introduced and derived and then combined to the describing differential equations.

Esslingen, Germany
April 2024

Andreas Öchsner

# Contents

1   **Fundamentals of Continuum Mechanical Modeling** .................   1
     1.1    Basic Equations for Structural Members ........................   1
     1.2    Beams and Plates ..........................................   4
     References ....................................................   6

2   **Thin Beams and Plates** .........................................   7
     2.1    Introduction ..............................................   7
     2.2    Kinematics Equations .......................................   7
     2.3    Constitutive Equations .....................................   12
     2.4    Equilibrium Equations ......................................   14
     2.5    Differential Equations .....................................   21
     References ....................................................   24

3   **Thick Beams and Plates** ........................................   25
     3.1    Introduction ..............................................   25
     3.2    Kinematics Equations .......................................   27
     3.3    Constitutive Equations .....................................   32
     3.4    Equilibrium Equations ......................................   36
     3.5    Differential Equations .....................................   39
     3.6    Alternative Formulation ....................................   41
     References ....................................................   47

4   **Comparison of the Approaches** ..................................   49
     References ....................................................   51

**Appendix A: Bending of Beams in the $xy$-Plane** .......................   53

**Index** ...........................................................   57

# Symbols and Abbreviations

## Latin Symbols (Capital Letters)

| | |
|---|---|
| $A$ | Area |
| $A$ | Shear area |
| $C$ | Scaler version of elasticity matrix |
| $\boldsymbol{C}$ | Elasticity matrix |
| $D$ | Abbreviation for bending stiffness $(D = EI_y)$, diameter |
| $D_b$ | Bending rigidity of a plate $(D = \frac{Eh^3}{12(1-\nu^2)})$ |
| $D_s$ | Normalized shear rigidity $(D_s = k_s Gh)$ |
| $\boldsymbol{D}$ | Elastic compliance matrix, generalized elasticity matrix of thick beam or plate |
| $\boldsymbol{D}_b$ | Thin plate elasticity matrix $(\boldsymbol{D} = \frac{h^3}{12}\boldsymbol{C})$ |
| $\boldsymbol{D}_s$ | Normalized shear rigidity matrix |
| $E$ | Modulus of elasticity (Young's modulus) |
| $EI$ | Bending stiffness of a beam |
| $F$ | Force |
| $G$ | Shear modulus |
| $I$ | Second moment of area |
| $L$ | Length |
| $M$ | Moment |
| $M^n$ | Normalized moment |
| $\boldsymbol{M}^n$ | Column matrix of normalized moments |
| $Q$ | Shear force |
| $Q^n$ | Normalized shear force |
| $\boldsymbol{Q}^n$ | Column matrix of normalized shear forces |
| $R$ | Radius |

## Latin Symbols (Small Letters)

| | |
|---|---|
| $a$ | Geometrical dimension (plate) |
| $b$ | Geometrical dimension (plate), distributed load |
| $\boldsymbol{b}$ | Column matrix of distributed loads |
| $e$ | Generalized strain |
| $\boldsymbol{e}$ | Column matrix of generalized strains |
| $h$ | Plate thickness |
| $k_{\mathrm{s}}$ | Shear correction factor |
| $m$ | Distributed moment |
| $q$ | Distributed force |
| $s$ | Generalized stress |
| $\boldsymbol{s}$ | Column matrix of generalized stresses |
| $t$ | Dimension |
| $u$ | Displacement |
| $\boldsymbol{u}$ | Column matrix of deformations |
| $x$ | Cartesian coordinate |
| $y$ | Cartesian coordinate |
| $z$ | Cartesian coordinate |

## Greek Symbols (Small Letters)

| | |
|---|---|
| $\gamma$ | Shear strain |
| $\varepsilon$ | Normal strain |
| $\boldsymbol{\varepsilon}$ | Strain column matrix |
| $\kappa$ | Curvature |
| $\boldsymbol{\kappa}$ | Curvature column matrix |
| $\nu$ | Poisson's ratio |
| $\sigma$ | Normal stress |
| $\boldsymbol{\sigma}$ | Stress column matrix |
| $\tau$ | Shear stress |
| $\phi$ | Rotational angle (thick beam) |
| $\varphi$ | Rotational angle (thin beam) |

## Mathematical Symbols

| | |
|---|---|
| $[\ldots]^{\mathrm{T}}$ | Transpose |
| $\mathcal{L}(\ldots)$ | Operator symbol |
| $\mathcal{L}_1(\ldots)$ | First order derivative symbol |
| $\mathcal{L}_2(\ldots)$ | Second order derivative symbol |

| | |
|---|---|
| $\mathcal{L}$ | Matrix of differential operators |
| $\mathcal{L}_1$ | Matrix of first order differential operators |
| $\mathcal{L}_2$ | Matrix of second order differential operators |
| $\partial\ldots$ | Partial derivative symbol (rounded $d$) |
| sin | Sine function |
| tan | Tangent function |

# Indices, Superscripted

| | |
|---|---|
| $\ldots^*$ | Alternative formulation |
| $\ldots^n$ | Normalized |

# Indices, Subscripted

| | |
|---|---|
| $\ldots_b$ | Bending |
| $\ldots_s$ | Shear |

# Abbreviations

| | |
|---|---|
| 1D | One-Dimensional |
| 2D | Two-Dimensional |
| BEM | Boundary Element Method |
| const. | Constant |
| FDM | Finite Difference Method |
| FEM | Finite Element Method |
| FVM | Finite Volume Method |
| PDE | Partial Differential Equation |

# Chapter 1
# Fundamentals of Continuum Mechanical Modeling

**Abstract** The first chapter introduces the major idea and the continuum mechanical background to model structural members. It is explained that mechanical members are described based on differential equations. These equations are obtained by combining the three basic equations of continuum mechanics, i.e., the kinematics relationship, the constitutive law, and the equilibrium equation. The second part introduces the one- and two-dimensional bending members, i.e., beams and plates. Furthermore, some explanations on the choice of the coordinate system for bending problems are provided.

## 1.1 Basic Equations for Structural Members

Let us first introduce some fundamental quantities, namely the concept of stress and strain and the relationship between these quantities. In the following we will only consider the simplest formulation, i.e., the so-called engineering quantities.

Consider an arbitrary force $F$ which is acting on a surface $A$, see Fig. 1.1. We can split this force into a component perpendicular to the surface $(F_\perp)$, and a component acting along this surface $(F_\parallel)$. This split allows us to define the normal stress, i.e.,

$$\sigma = \frac{F_\perp}{A}, \qquad (1.1)$$

as well as the shear stress, i.e.,

$$\tau = \frac{F_\parallel}{A}. \qquad (1.2)$$

In the context of the engineering definition used here, it is assumed that the surface area $(A)$ remains unchanged during the deformation. These two stress quantities are a typical measure for the internal loading of the material.

Let us now look at the deformation of the material, see Fig. 1.2. Assuming a tensile test situation as shown in Fig. 1.2a, the (engineering) normal strain is defined as the

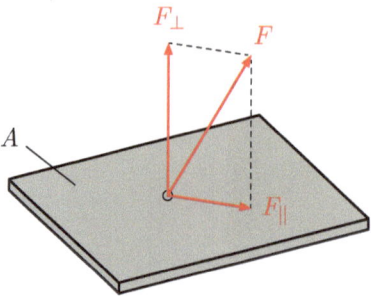

**Fig. 1.1** Definition of normal and shear stress

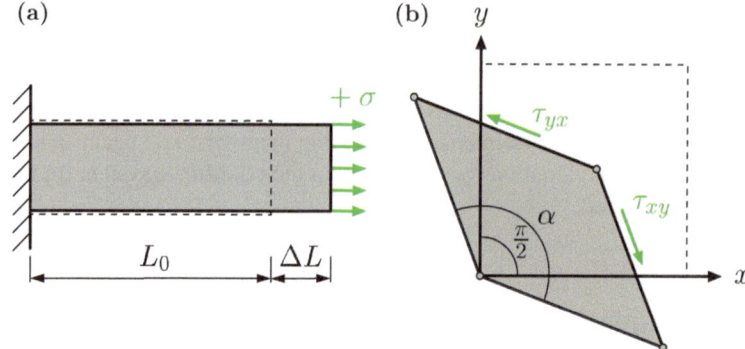

**Fig. 1.2** Definition of **(a)** normal and **(b)** shear strain

change in length $\Delta L$ of two adjacent points relative to the initial length $L_0$ in the unloaded state:

$$\varepsilon = \frac{\Delta L}{L_0}. \tag{1.3}$$

From this equation it can be seen that the normal strain is dimensionless. The term (engineering) shear strain refers to the change in angle of an originally rectangular angular element, see Fig. 1.2:

$$\gamma = \alpha - \frac{\pi}{2}. \tag{1.4}$$

The shear strain is given in radians (rad), and is therefore dimensionless. The quantities for the loading of the material (stresses) and for the deformation (strains) are related to each other by the constitutive equation (in the simplest case by Hooke's law). For a pure tensile load case, one-dimensional Hooke's law is written as

$$\sigma = E\varepsilon, \tag{1.5}$$

where the constant of proportionality ($E$) is called Young's modulus. Alternatively, for a pure shear stress state, one-dimensional Hooke's law is written as

$$\tau = G\gamma, \tag{1.6}$$

where the constant of proportionality $(G)$ is called shear modulus.

Moving to two- or three-dimensional stress states, Hooke's law for isotropic materials requires a set of two independent material properties. In the context of mechanical engineering, this can be the Young's modulus and Poisson's ratio. It should be noted here these two material parameters can be determined in a uniaxial tensile test using a universal testing machine.

Modeling of structural members follows a common approach from continuum mechanics [1], see Fig. 1.3.

A combination of the kinematics equation (i.e., the relation between the strains and deformations) with the constitutive equation (i.e., the relation between the stresses and strains) and the equilibrium equation (i.e., the equilibrium between the internal reactions and the external loads) results in a partial differential equation, or a corresponding system of differential equations. Limited to simple problems and config-

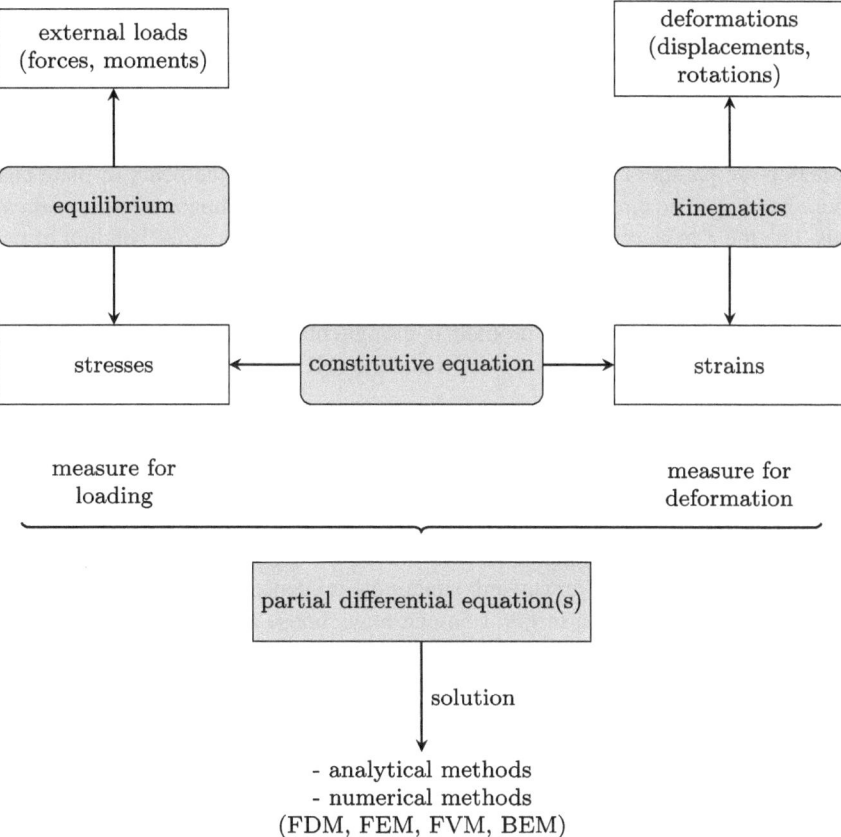

**Fig. 1.3** Continuum mechanical modeling of structural members

urations, *analytical* solutions are possible. These analytical solutions are then exact in the frame of the assumptions made. However, the solution of complex problems requires the application of numerical methods such as the finite difference method (see [6, 9, 12]), the finite element method (see [2, 10, 13]), the finite volume method (see [14]), or the boundary element method (see [4]). These *numerical* solutions are, in general, no longer exact since the numerical methods provide only an approximate solution. Thus, the major task of an engineer is then to ensure that the approximate solution is as good as possible and close to reality. This requires a lot of experience and a solid foundation in the basics of applied mechanics, materials science, mathematics, and even programming.

## 1.2  Beams and Plates

Beams and plates are structural members that deform perpendicular to the center line, i.e., the $x$-axis for a beam (see Fig. 1.4a) or the center plane, i.e., the $xy$-plane for a plate (see Fig. 1.4b). The deformation is characterized by a translation in the $z$-direction and one (in the case of a beam for symmetrical bending) or two (in the case of a plate) rotations. The external loads, which are considered within the following description, are single forces $F_z$, single moments $M$, distributed forces $q_z$, and distributed moments $m$. The external forces have in common that their lines of action are parallel to the vertical axis, i.e., the $z$-axis, of the beam or plate and cause bending. The direction of the momentum vectors is orthogonal to the vertical axis, i.e., the $z$-axis, of the beam or plate and these moments cause bending of the structural member. This is a different type of deformation compared to the 1D bar or 2D plane elasticity element [13].

To describe a beam bending problem in a single plane, there are different options on how to choose the coordinate system and the orientation of the single axes, see Fig. 1.5.

Let us have the beam in a horizontal position and the deflection should occur in the vertical direction. A choice common to mathematical education in secondary school would be to have the $x$-axis aligned with the horizontal longitudinal axis of the beam[1] and the $y$-axis vertically upwards, see Fig. 1.5a. Civil engineers would rather prefer to have the $y$-axis vertically downwards, see Fig. 1.5b. This would give a positive deflection (downwards) under the influence of gravity. Alternatively to the configuration shown in Fig. 1.5a one may choose a $xz$-plane, see Fig. 1.5c. This choice has advantages if similarities between a bending beam and a bending plate should be shown since the thickness coordinate for two-dimensional structural members is commonly the $z$-coordinate. Obviously, the $z$-coordinate may be oriented downwards, see Fig. 1.5d. Nevertheless, an engineer must be able to master both

---

[1] The common approach in analytical mechanics is to align the $x$-axis with the longitudinal axis of the beam. However, in the context of the finite element method, the local $z$-axis might be oriented along the element, see [10].

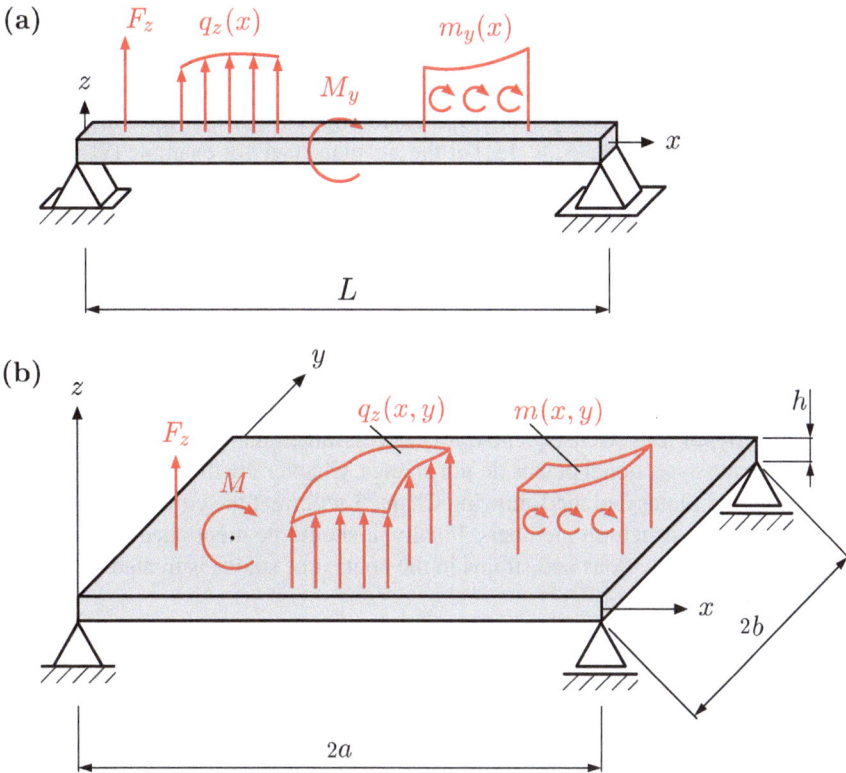

**Fig. 1.4** General configurations of **a** a beam (bending in the $xz$-plane) and **b** a plate problem

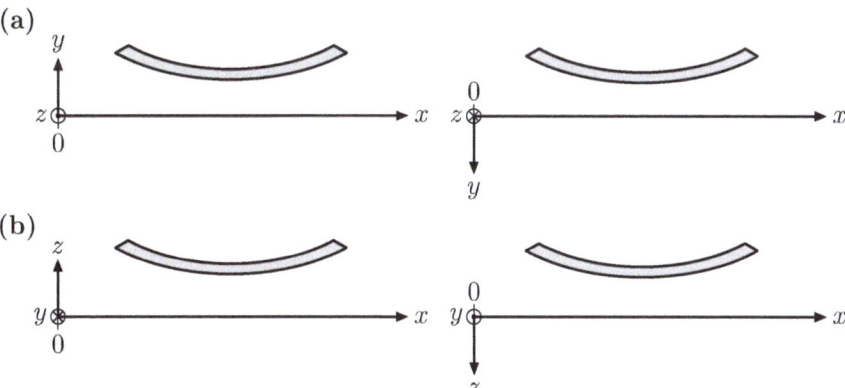

**Fig. 1.5** Different ways to introduce beam bending problems in a single plane: **a** $xy$-plane and **b** $xz$-plane

configurations, i.e., bending in the $xy$-plane and the $xz$-plane since this relates to the general case of unsymmetrical bending [11]. The above mentioned options on how to introduce the coordinate system for single-plane bending problems is also reflected in common teaching books where bending is either first introduced in the $xy$-plane (see for example [3, 8, 15]) or the $xz$-plane (see for example [7]). Due to the similarity with the two-dimensional approach, the $xz$-plane will be used in the following chapters to describe the bending behavior of beams in a single plane.

The classic theories of bending distinguish between shear-rigid and shear-flexible models [5]. The shear-rigid models neglect the shear deformation from the shear forces. This theory implies that a cross-sectional plane which was perpendicular to horizontal axis before the deformation remains in the deformed state perpendicular to the horizontal axis. Furthermore, it is assumed that a cross-sectional plane stays plane and unwarped in the deformed state. These two assumptions are also known as Bernoulli's hypothesis. Consequently, it is also assumed that the geometric dimensions of the cross-sectional planes do not change. Chapter 2 will refer to shear-rigid thin beams and plates and the following Chapt. 3 will treat the corresponding thick versions of these structural members. Finally, it should be mentioned that only the case of small deformations and strains in the context of statics is treated in the following chapters.

# References

1. Altenbach H, Öchsner A (eds) (2020) Encyclopedia of continuum mechanics. Springer, Berlin
2. Bathe K-J (1996) Finite element procedures. Prentice-Hall, Upper Saddle River
3. Beer FP, Johnston ER Jr, DeWolf JT, Mazurek DF (2009) Mechanics of materials. McGraw-Hill, New York
4. Banerjee PK (1994) Boundary element methods in engineering. McGraw-Hill, London
5. Blaauwendraad J (2010) Plates and FEM: surprises and pitfalls. Springer, Dordrecht
6. Forsythe GE, Wasow WR (1960) Finite-difference methods for partial differential equations. Wiley, New York
7. Gross D, Hauger W, Schröder J, Wall WA, Bonet J (2011) Engineering mechanics 2: mechanics of materials. Springer, Berlin
8. Hibbeler RC (2011) Mechanics of materials. Prentice Hall, Singapore
9. Mitchell AR, Griffiths DF (1980) The finite difference method in partial differential equations. Wiley, New York
10. Öchsner A, Öchsner M (2018) A first introduction to the finite element analysis program MSC Marc/Mentat. Springer, Cham
11. Öchsner A (2021) Classical beam theories of structural mechanics. Springer, Cham
12. Öchsner A (2021) Structural mechanics with a pen: a guide to solve finite difference problems. Springer, Cham
13. Öchsner A (2023) Computational statics and dynamics: an introduction based on the finite element method. Springer, Singapore
14. Petrova R (ed) (2012) Finite volume method- powerful means of engineering design. InTech, Rijeka
15. Popov L (1990) Engineering mechanics of solids. Prentice-Hall, Englewood Cliffs

# Chapter 2
# Thin Beams and Plates

**Abstract** This chapter presents the analytical description of thin beam and plate members. Based on the three basic equations of continuum mechanics, i.e., the kinematics relationship, the constitutive law and the equilibrium equation, the partial differential equations, which describe the physical problems, are derived.

## 2.1  Introduction

Schematic representations of a thin beam and plate are shown in Fig. 2.1. *Thin* means in the case of a beam that the length $L$ is approximately ten times a characteristic dimension of the cross section, e.g. the width $b$ and/or the height $h$, see Fig. 2.1a. Similarly, thin in the case of a plate means[1] that the in-plane dimensions $2a$ and $2b$ are approximately ten times the thickness $h$ of the plate, see Fig. 2.1b.

## 2.2  Kinematics Equations

The kinematics or strain-displacement relations extract the strain field contained in a displacement field. Let us first derive the kinematics relation for a thin beam element, see Fig. 2.2a, see [2]. It is assumed that a beam of length $L$ is under constant moment loading $M_y(x) = \text{const.} > 0$, meaning under *pure* bending. Two conventions, which are applied here for the beam as well as the plate element, should be highlighted here:

- A rotational angle $\varphi$ is positive if the vector of the rotational direction is pointing in the corresponding positive axis direction.
- The internal moments (i.e., the stress resultants) are taken to be positive if they cause a tensile stress (positive) at a point with positive z-coordinate.

---

[1] The thin plate theory is also connected with the name Kirchhoff.

© The Author(s), under exclusive license to Springer Nature Switzerland AG 2025
A. Öchsner, *The Fundamental Equations of Beams and Plates*,
SpringerBriefs in Continuum Mechanics,
https://doi.org/10.1007/978-3-031-76276-5_2

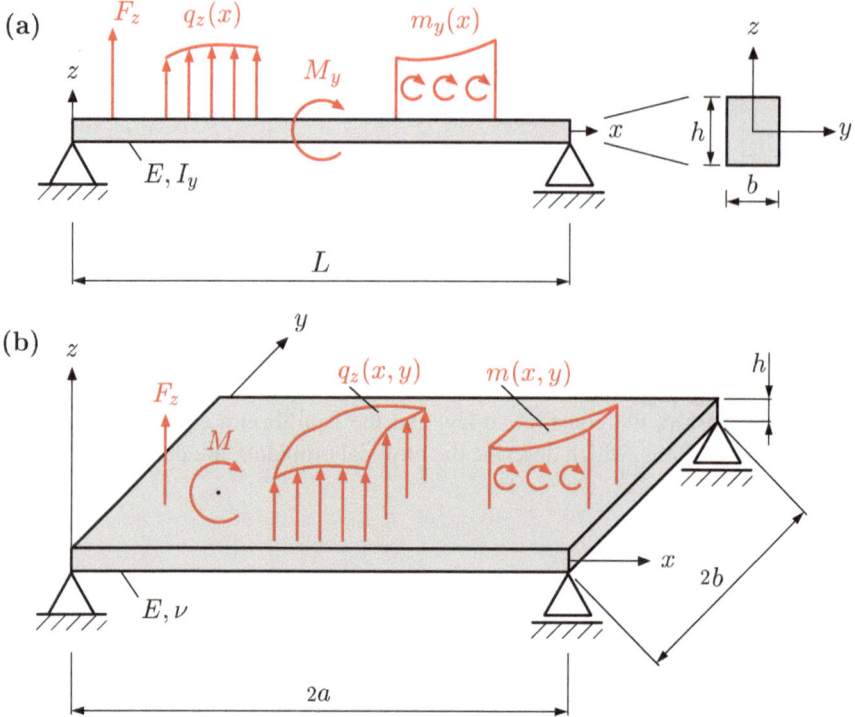

**Fig. 2.1** General configurations of **a** a thin beam (bending in the $xz$-plane) and **b** a thin plate problem

From the relation of the right-angled triangle $0'1'2'$ in Fig. 2.2a, this means[2] $\sin \varphi_y = \frac{u_x}{z}$, the following relation results for small angles ($\sin \varphi_y \approx \varphi_y$):

$$u_x = +z\varphi_y . \tag{2.1}$$

Furthermore, it holds that the rotation angle equals the slope of the center line for small angles:

$$\tan \varphi_y = \frac{-\,\mathrm{d}u_z(x)}{\mathrm{d}x} \approx \varphi_y . \tag{2.2}$$

If Eqs. (2.2) and (2.1) are combined, the following results:

$$u_x = -z\frac{\mathrm{d}u_z(x)}{\mathrm{d}x} . \tag{2.3}$$

---

[2] Note that according to the assumptions of the thin beam the lengths $01$ and $0'1'$ remain unchanged.

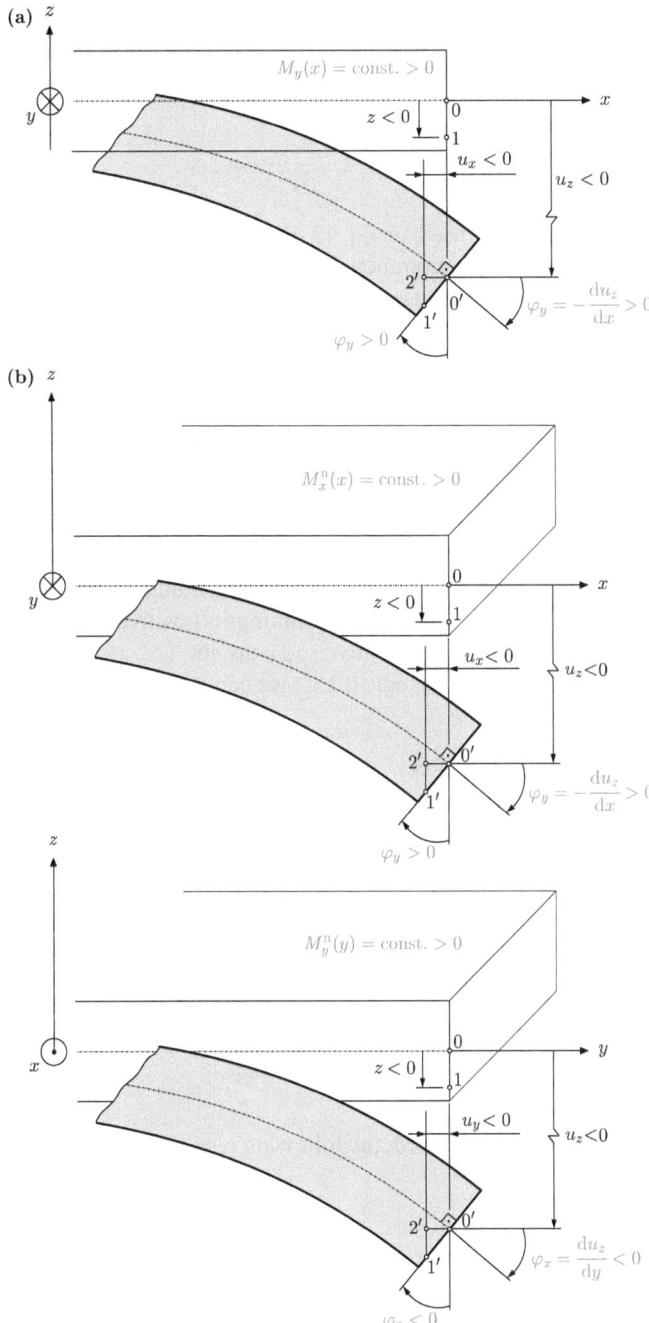

**Fig. 2.2** Configurations for the derivation of kinematics relations: **a** a thin beam (bending in the $xz$-plane) and **b** a thin plate problem

From the last equation, using the classical engineering definition of normal strain [1, 4], i.e., $\varepsilon_x = \frac{du_x}{dx}$, one obtains the kinematics relationship by differentiation with respect to the $x$-coordinate as:

$$\varepsilon_x(x, z) = -z\frac{d^2 u_z(x)}{dx^2} = z\kappa_x .\tag{2.4}$$

It should be noted that we used in Eq. (2.4) also the definition of the so-called curvature $\kappa_x$. If we replace the common formulation of the second order derivative, i.e. $\frac{d^2 \cdots}{dx^2}$, by a formal operator symbol, i.e. $\mathcal{L}_2(\cdots)$, the basic kinematics equation can be stated in a more formal way as

$$\varepsilon_x(x, z) = -z\mathcal{L}_2 (u_z(x)) .\tag{2.5}$$

Let us now derive the kinematics relation for a thin plate element. This relates in the $xz$-plane the variation of $u_x$ across the plate thickness in terms of the displacement $u_z$. For this purpose, let us first imagine that a plate element is bent around the $y$-axis, see Fig. 2.2b (top image). We assume in the following the same definition of the rotational angle $\varphi_y$ as for the thin beam. This means that the angle $\varphi_y$ is positive if the vector of the rotational direction is pointing in positive $y$-axis. The internal bending moment is assumed to be positive and constant, i.e., $M_x^{n}(x) = \text{const.} > 0$.

Looking at the right-angled triangle $0'1'2'$, we can state that[3]

$$\sin(\varphi_y) = \frac{\overline{2'0'}}{\overline{0'1'}} = \frac{-u_x}{-z},\tag{2.6}$$

which results for small angles $(\sin(\varphi_y) \approx \varphi_y)$ in:

$$u_x = +z\varphi_y.\tag{2.7}$$

Looking at the curved center line in Fig. 2.2b (top image), it holds that the slope of the tangent line at $0'$ equals:

$$\tan(\varphi_y) = \frac{-du_z}{dx} \approx \varphi_y .\tag{2.8}$$

If Eqs. (2.7) and (2.8) are combined, the following results:

$$u_x = -z\frac{du_z}{dx} .\tag{2.9}$$

Considering a plate which is bent around the $x$-axis (see Fig. 2.2b (bottom image)) and following the same line of reasoning (the angle $\varphi_x$ is assumed positive if the

---

[3] Note that according to the assumptions of the classical *thin* plate theory the lengths $\overline{01}$ and $\overline{0'1'}$ remain unchanged.

vector of the rotational direction is pointing in positive $x$-axis), similar equations can be derived for $u_y$:

$$-\varphi_x \approx \frac{-\mathrm{d}u_z}{\mathrm{d}y},\tag{2.10}$$

$$u_y = -z\varphi_x,\tag{2.11}$$

$$u_y = -z\frac{\mathrm{d}u_z}{\mathrm{d}y}.\tag{2.12}$$

One may find in the scholarly literature other definitions of the rotational angles [1, 3, 5, 6]. The angle $\varphi_y$ is introduced in the $xz$-plane (see Fig. 2.2b (top image)) whereas $\varphi_x$ is introduced in the $yz$-plane (see Fig. 2.2b (bottom image)). These definitions are closer to the classical definitions of the angles in the scope of finite elements but not conform with the definitions of the stress resultants (see $M_x^n$ and $M_y^n$ in Fig. 2.4b). Other definitions assume, for example, that the rotational angle $\varphi_x$ (now defined in the $xz$-plane) is positive if it leads to a positive displacement $u_x$ at the positive $z$-side of the neutral axis. The same definition holds for the angle $\varphi_y$ (now defined in the $yz$-plane).

Using classical engineering definitions of strain, the following relations can be obtained [1, 4]:

$$\varepsilon_x = \frac{\partial u_x}{\partial x} \overset{(2.9)}{=} \frac{\partial}{\partial x}\left(-z\frac{\partial u_z}{\partial x}\right) = -z\frac{\partial^2 u_z}{\partial x^2} = z\kappa_x,\tag{2.13}$$

$$\varepsilon_y = \frac{\partial u_y}{\partial y} \overset{(2.12)}{=} \frac{\partial}{\partial y}\left(-z\frac{\partial u_z}{\partial y}\right) = -z\frac{\partial^2 u_z}{\partial y^2} = z\kappa_y,\tag{2.14}$$

$$\gamma_{xy} = \frac{\partial u_x}{\partial y} + \frac{\partial u_y}{\partial x} \overset{(2.9),(2.12)}{=} -2z\frac{\partial^2 u_z}{\partial x\partial y} = z\kappa_{xy}.\tag{2.15}$$

In matrix notation, these three relationships for the strains can be written as

$$\begin{bmatrix} \varepsilon_x \\ \varepsilon_y \\ \gamma_{xy} \end{bmatrix} = -z\begin{bmatrix} \frac{\partial^2}{\partial x^2} \\ \frac{\partial^2}{\partial y^2} \\ \frac{2\partial^2}{\partial x\partial y} \end{bmatrix} u_z = z\begin{bmatrix} \kappa_x \\ \kappa_y \\ \kappa_{xy} \end{bmatrix},\tag{2.16}$$

or symbolically as

$$\boldsymbol{\varepsilon} = -z\mathcal{L}_2 u_z = z\boldsymbol{\kappa}.\tag{2.17}$$

**Table 2.1** Comparison of the kinematics equations for thin beam (bending in the $xz$-plane) and thin plate elements

| Thin beam | Thin plate |
|---|---|
| Specific formulation | |
| $\varepsilon_x(x, z) = -z\dfrac{\mathrm{d}^2 u_z(x)}{\mathrm{d}x^2} = z\kappa_x$ | $\begin{bmatrix} \varepsilon_x \\ \varepsilon_y \\ \gamma_{xy} \end{bmatrix} = -z \begin{bmatrix} \frac{\partial^2}{\partial x^2} \\ \frac{\partial^2}{\partial y^2} \\ \frac{2\partial^2}{\partial x \partial y} \end{bmatrix} u_z = z \begin{bmatrix} \kappa_x \\ \kappa_y \\ \kappa_{xy} \end{bmatrix}$ |
| General formulation | |
| $\varepsilon_x(x, z) = -z\mathcal{L}_2\left(u_z(x)\right) = z\kappa_x$ | $\boldsymbol{\varepsilon}(x, y, z) = -z\mathcal{L}_2 u_z = z\boldsymbol{\kappa}$ |

Let us summarize at the end of this section the kinematics relations for thin beams and thin plates, see Table 2.1. The similarity is particularly visible for the general formulation.

## 2.3  Constitutive Equations

The one-dimensional Hooke's law can be used in the case of the thin beam, since, according to the assumptions, only normal stresses (see Fig. 2.3a) are regarded in this section:

$$\sigma_x = E\varepsilon_x, \tag{2.18}$$

where the material parameter $(E)$ is the so-called Young's modulus or modulus of elasticity. In a more general way, one may write Eq. (2.18) with the elasticity matrix, or more precisely in this one-dimensional case with the scaler version $(C)$ as:

$$\sigma_x = C\varepsilon_x. \tag{2.19}$$

The classical plate theory assumes a plane stress state (see Fig. 2.3b), i.e., a two-dimensional stress state, and the constitutive equation can be stated as:

$$\begin{bmatrix} \sigma_x \\ \sigma_y \\ \tau_{xy} \end{bmatrix} = \frac{E}{1-\nu^2} \begin{bmatrix} 1 & \nu & 0 \\ \nu & 1 & 0 \\ 0 & 0 & \frac{1-\nu}{2} \end{bmatrix} \begin{bmatrix} \varepsilon_x \\ \varepsilon_y \\ \gamma_{xy} \end{bmatrix}, \tag{2.20}$$

or rearranged for the elastic compliance form:

$$\begin{bmatrix} \varepsilon_x \\ \varepsilon_y \\ \gamma_{xy} \end{bmatrix} = \frac{1}{E} \begin{bmatrix} 1 & -\nu & 0 \\ -\nu & 1 & 0 \\ 0 & 0 & 2(\nu+1) \end{bmatrix} \begin{bmatrix} \sigma_x \\ \sigma_y \\ \tau_{xy} \end{bmatrix}, \tag{2.21}$$

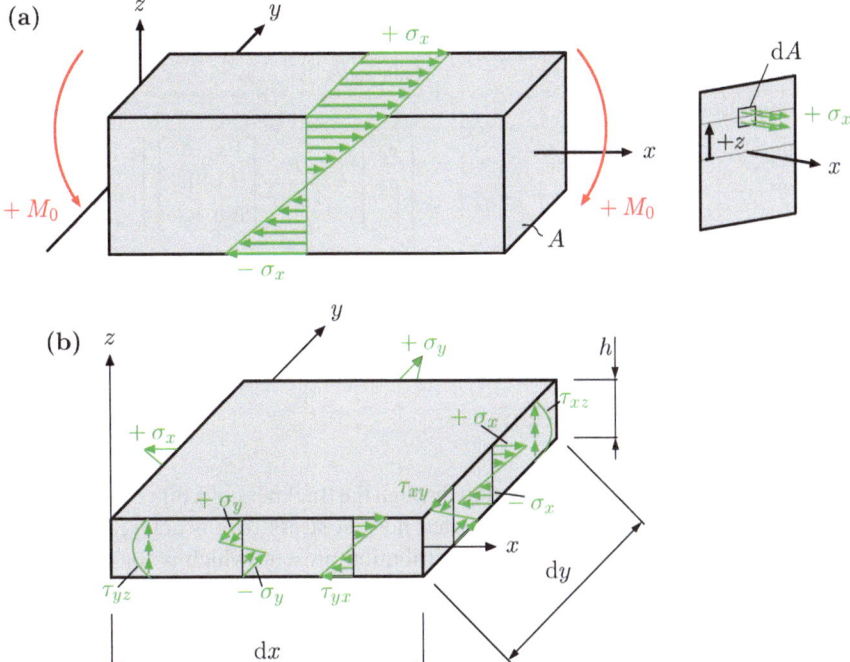

**Fig. 2.3** Stresses acting on **a** a thin beam (bending in the $xz$-plane) and **b** a thin plate

where we have now in addition to Young's modulus ($E$) the so-called Poisson's ratio ($\nu$) as a material parameter. The last two equations can be written in matrix notation as

$$\boldsymbol{\sigma} = \boldsymbol{C}\boldsymbol{\varepsilon}, \tag{2.22}$$

or

$$\boldsymbol{\varepsilon} = \boldsymbol{D}\boldsymbol{\sigma}, \tag{2.23}$$

where $\boldsymbol{C}$ is the elasticity matrix and $\boldsymbol{D} = \boldsymbol{C}^{-1}$ is the elastic compliance matrix.

Let us summarize at the end of this section the constitutive equations for thin beams and thin plates, see Table 2.2. The similarity is particularly visible for the general formulation.

**Table 2.2** Comparison of the constitutive equations for thin beam (bending in the $xz$-plane) and thin plate elements

| Thin beam | Thin plate |
|---|---|
| Specific formulation | |
| $\sigma_x = E\varepsilon_x$ | $\begin{bmatrix} \sigma_x \\ \sigma_y \\ \tau_{xy} \end{bmatrix} = \dfrac{E}{1-\nu^2} \begin{bmatrix} 1 & \nu & 0 \\ \nu & 1 & 0 \\ 0 & 0 & \frac{1-\nu}{2} \end{bmatrix} \begin{bmatrix} \varepsilon_x \\ \varepsilon_y \\ \gamma_{xy} \end{bmatrix}$ |
| General formulation | |
| $\sigma_x = C\varepsilon_x$ | $\boldsymbol{\sigma} = \boldsymbol{C}\boldsymbol{\varepsilon}$ |

## 2.4  Equilibrium Equations

Let us first look at the stress distribution through the thickness of a thin beam element as shown in Fig. 2.3a. A linear distributed normal stress ($\sigma_x$) is acting. For many load cases, there is also a shear stress distribution present, which is for thin beams disregarded compared to the normal stress distribution. The action of these stresses can be represented by the so-called stress resultants, i.e. the internal bending moment ($M_y$) and the internal shear force ($Q_z$) defined as:

$$M_y(x) = \int_A z\sigma_x \mathrm{d}A, \tag{2.24}$$

$$Q_z(x) = \int_A \tau_{xz} \mathrm{d}A. \tag{2.25}$$

The equilibrium conditions are derived from an infinitesimal beam element of length $\mathrm{d}x$, which is loaded by a constant distributed force $q_z$, see Fig. 2.4a. The internal reactions are drawn on both cut faces, i.e. at locations $x$ and $x + \mathrm{d}x$. One can see that a positive shear force is oriented in the positive $z$-direction at the right-hand face[4] and that a positive bending moment has the same rotational direction as the positive $y$-axis (right-hand grip rule[5]). The orientation of shear force and bending moment is reversed at the left-hand face in order to cancel in sum the effect of the internal reactions at both faces. This convention for the direction of the internal reactions is maintained in the following for the thin beam. Furthermore, it can be derived from Fig. 2.4a that an upwards directed *external* force or alternatively a mathematically

---

[4] A positive cut face is defined by the surface normal on the cut plane which has the same orientation as the positive $x$-axis. It should be regarded that the surface normal is always directed outward.

[5] If the axis is grasped with the right hand in a way so that the spread out thumb points in the direction of the positive axis, the bent fingers then show the direction of the positive rotational direction.

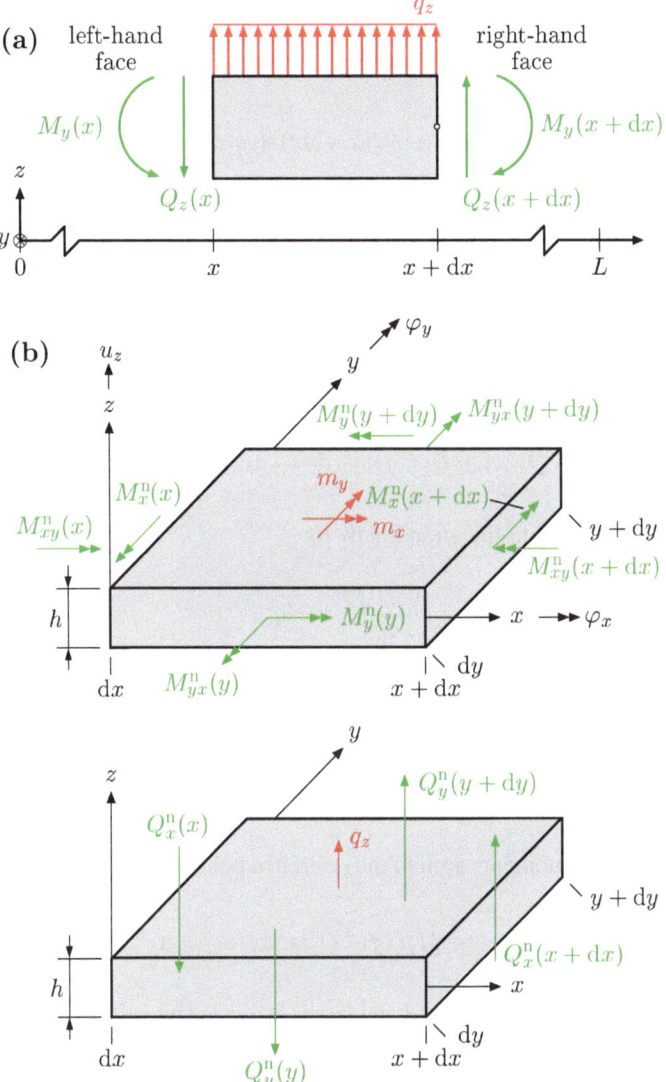

**Fig. 2.4** Infinitesimal elements with internal reactions: **a** a thin beam (bending in the $xz$-plane) and **b** a thin plate problem

positive oriented *external* moment at the right-hand face leads to a positive shear force or alternatively a positive internal moment. In a corresponding way, it results that a downwards directed *external* force or alternatively a mathematically negative oriented *external* moment at the left-hand face leads to a positive shear force or alternatively a positive internal moment.

The equilibrium condition will be determined in the following for the vertical forces. Assuming that forces in the direction of the positive $z$-axis are considered positive, the following results:

$$- Q_z(x) + Q_z(x + \mathrm{d}x) + q_z \mathrm{d}x = 0 . \tag{2.26}$$

If the shear force on the right-hand face is expanded in a Taylor's series of first order, meaning a linearization of the function as

$$Q_z(x + \mathrm{d}x) \approx Q_z(x) + \frac{\mathrm{d}Q_z(x)}{\mathrm{d}x} \mathrm{d}x , \tag{2.27}$$

Equation (2.26) results in

$$- Q_z(x) + Q_z(x) + \frac{\mathrm{d}Q_z(x)}{\mathrm{d}x} \mathrm{d}x + q_z \mathrm{d}x = 0 , \tag{2.28}$$

or alternatively after simplification finally to:

$$\frac{\mathrm{d}Q_z(x)}{\mathrm{d}x} = -q_z . \tag{2.29}$$

For the special case that no distributed load is acting ($q_z = 0$), Eq. (2.29) simplifies to:

$$\frac{\mathrm{d}Q_z(x)}{\mathrm{d}x} = 0 . \tag{2.30}$$

The equilibrium of moments around the reference point at $x + \mathrm{d}x$ gives:

$$M_y(x + \mathrm{d}x) - M_y(x) - Q_z(x)\mathrm{d}x + \frac{1}{2} q_z \mathrm{d}x^2 = 0 . \tag{2.31}$$

If the bending moment on the right-hand face is expanded into a Taylor's series of first order similar to Eq. (2.27) and consideration that the term $\frac{1}{2} q_z \mathrm{d}x^2$ as infinitesimal small size of higher order can be disregarded, finally the following results:

$$\frac{\mathrm{d}M_y(x)}{\mathrm{d}x} = Q_z(x) . \tag{2.32}$$

The combination of Eqs. (2.29) and (2.32) leads to the relationship between the bending moment and the distributed load:

$$\frac{\mathrm{d}^2 M_y(x)}{\mathrm{d}x^2} = \frac{\mathrm{d}Q_z(x)}{\mathrm{d}x} = -q_z(x). \tag{2.33}$$

Let us first look at the stress distributions through the thickness of a classical plate element $dxdyh$ as shown in Fig. 2.3b. Linear distributed normal stresses ($\sigma_x$, $\sigma_y$), linear distributed shear stresses ($\tau_{yx}$, $\tau_{xy}$), and parabolic distributed shear stresses ($\tau_{yz}$, $\tau_{xz}$) can be identified. The action of these stresses can be represented by the so-called stress resultants, i.e. bending moments and shear forces as shown in Fig. 2.4b. These stress resultants, i.e. the internal bending moments, are taken to be positive if they cause a tensile stress (positive) at a point with positive $z$-coordinate[6]. Looking at the bending of a plate in the $xz$-plane (see 2.5b (top)), it can be seen that the internal reactions have the same orientation as in the case of the corresponding beam problem. Only the indices of the bending moment and shear force are differently assigned. Attention should be paid to the plate deformation in the $yz$-plane (see 2.5b (bottom)). Here, the internal bending moment rotates at the positive face opposite to the rotational direction as the positive $x$-axis in order to produce a positive tensile stress at a point with positive $z$-coordinate.

To better illustrate the different conventions for the internal reactions depending on the structural element and the bending plane, Fig. 2.5 represents orthogonal views on the structural members.

The stress resultants are obtained as in the case of beams[7] by integrating over the stress distributions. In the case of plates, however, the integration is only performed over the thickness, i.e. the moments and forces are given per unit length (normalized with the corresponding side length of the plate element). The normalized (superscript 'n') bending moments are obtained as:

$$M_x^n = \frac{M_x}{dy} = \int_{-h/2}^{h/2} z\sigma_x dz, \tag{2.34}$$

$$M_y^n = \frac{M_y}{dx} = \int_{-h/2}^{h/2} z\sigma_y dz. \tag{2.35}$$

The twisting moment per unit length reads:

$$M_{xy}^n = M_{yx}^n = \frac{M_{xy}}{dy} = \frac{M_{yx}}{dx} = \int_{-h/2}^{h/2} z\tau_{xy} dz. \tag{2.36}$$

---

[6] This definition includes the definition of a positive bending moment for beam bending in the $xz$-plane, see Fig. 2.5a. Here, we used the definition that a positive bending moment has the same rotational direction as the positive $y$-axis.

[7] See Eq. (2.24) for the thin beam.

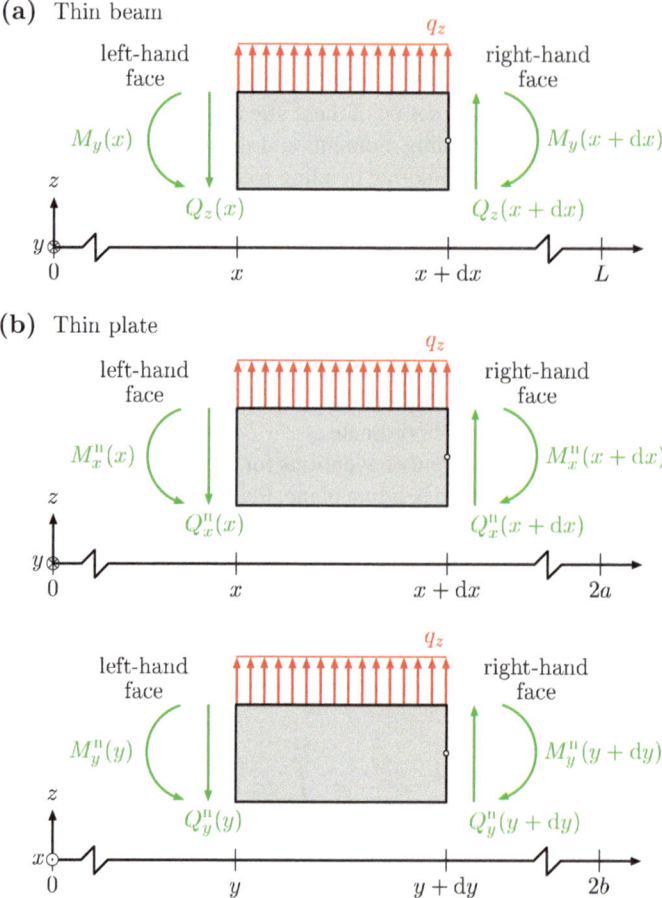

**Fig. 2.5** Alternative representation of infinitesimal elements with internal reactions: **a** a thin beam (bending in the $xz$-plane) and **b** a thin plate problem (orthogonal views on $xz$-plane (top) and $yz$-plane (bottom))

Furthermore, the shear forces per unit length are calculated in the following way:

$$Q_x^{\mathrm{n}} = \frac{Q_x}{\mathrm{d}y} = \int\limits_{-h/2}^{h/2} \tau_{xz}\mathrm{d}z, \tag{2.37}$$

$$Q_y^{\mathrm{n}} = \frac{Q_y}{\mathrm{d}x} = \int\limits_{-h/2}^{h/2} \tau_{yz}\mathrm{d}z. \tag{2.38}$$

It should be noted that a slightly different notation when compared to the beam problems is used here. The bending moment around the $y$-axis is now called $M_x^n$ (which directly corresponds to the causing stress $\sigma_x$) while in the beam notation it was $M_y$, see Fig. 2.4a. Nevertheless, the orientation remains the same. The shear force, which was in the case of the beams given as $Q_z$ is now either $Q_x^n$ or $Q_y^n$. Thus, in the case of this plate notation, the index refers rather to the plane (check the surface normal vector) in which the corresponding resultant (vector) is located.

The equilibrium condition will be determined in the following for the vertical forces. Assuming that the distributed force is constant ($q_z(x, y) \rightarrow q_z$) and that forces in the direction of the positive $z$-axis are considered positive, the following results:

$$- Q_x^n(x)\mathrm{d}y - Q_y^n(y)\mathrm{d}x + Q_x^n(x + \mathrm{d}x)\mathrm{d}y + Q_y^n(y + \mathrm{d}y)\mathrm{d}x + q_z\mathrm{d}x\mathrm{d}y = 0. \tag{2.39}$$

Evaluating the shear forces at $x + \mathrm{d}x$ and $y + \mathrm{d}y$ in a Taylor's series of first order, meaning

$$Q_x^n(x + \mathrm{d}x) \approx Q_x^n(x) + \frac{\partial Q_x^n}{\partial x}\mathrm{d}x, \tag{2.40}$$

$$Q_y^n(y + \mathrm{d}y) \approx Q_y^n(y) + \frac{\partial Q_y^n}{\partial y}\mathrm{d}y, \tag{2.41}$$

Equation (2.39) results in

$$\frac{\partial Q_x^n}{\partial x}\mathrm{d}x\mathrm{d}y + \frac{\partial Q_y^n}{\partial y}\mathrm{d}y\mathrm{d}x + q_z\mathrm{d}x\mathrm{d}y = 0, \tag{2.42}$$

or alternatively after simplification to:

$$\frac{\partial Q_x^n}{\partial x} + \frac{\partial Q_y^n}{\partial y} + q_z = 0. \tag{2.43}$$

The equilibrium of moments around the reference axis at $x + \mathrm{d}x$ (positive if the moment vector is pointing in positive $y$-axis) gives:

$$M_x^n(x + \mathrm{d}x)\mathrm{d}y - M_x^n(x)\mathrm{d}y + M_{yx}^n(y + \mathrm{d}y)\mathrm{d}x - M_{yx}^n\mathrm{d}x$$
$$- Q_y^n(y)\mathrm{d}x\tfrac{\mathrm{d}x}{2} + Q_y^n(y + \mathrm{d}y)\mathrm{d}x\tfrac{\mathrm{d}x}{2} - Q_x^n(x)\mathrm{d}y\mathrm{d}x + q_z\mathrm{d}x\mathrm{d}y\tfrac{\mathrm{d}x}{2} = 0. \tag{2.44}$$

Expanding the stress resultants at $x + \mathrm{d}x$ and $y + \mathrm{d}y$ into a Taylor's series of first order, meaning

$$M_x^n(x + \mathrm{d}x) = M_x^n(x) + \frac{\partial M_x^n}{\partial x}\mathrm{d}x, \tag{2.45}$$

$$M_{yx}^n(y + dy) = M_{yx}^n(y) + \frac{\partial M_{yx}^n}{\partial y} dy \,, \tag{2.46}$$

$$Q_y^n(y + dy) = Q_y^n(y) + \frac{\partial Q_y^n}{\partial y} dy \,, \tag{2.47}$$

Equation (2.44) results in

$$\frac{\partial M_x^n}{\partial x} dx dy + \frac{\partial M_{yx}^n}{\partial y} dy dx + \frac{\partial Q_y^n}{\partial y} dy dx \frac{dx}{2} - Q_x^n(x) dy dx + q_z dx dy \frac{dx}{2} = 0 \,. \tag{2.48}$$

Seeing that the terms of third order $(dx\,dy\,dx)$ are considered as infinitesimally small and because of $M_{yx}^n = M_{xy}^n$, finally the following results:

$$\frac{\partial M_x^n}{\partial x} + \frac{\partial M_{xy}^n}{\partial y} - Q_x^n = 0 \,. \tag{2.49}$$

In a similar way, the equilibrium of moments around the reference axis at $y + dy$ finally gives:

$$\frac{\partial M_y^n}{\partial y} + \frac{\partial M_{xy}^n}{\partial x} - Q_y^n = 0 \,. \tag{2.50}$$

Thus, the three equilibrium equations can be summarized as follows:

$$\frac{\partial Q_x^n}{\partial x} + \frac{\partial Q_y^n}{\partial y} + q_z = 0, \tag{2.51}$$

$$\frac{\partial M_x^n}{\partial x} + \frac{\partial M_{xy}^n}{\partial y} - Q_x^n = 0, \tag{2.52}$$

$$\frac{\partial M_y^n}{\partial y} + \frac{\partial M_{xy}^n}{\partial x} - Q_y^n = 0. \tag{2.53}$$

Rearranging Eqs. (2.52) and (2.53) for $Q^n$ and introducing in Eq. (2.51) finally gives the combined equilibrium equation as:

$$\frac{\partial^2 M_x^n}{\partial x^2} + 2\frac{\partial^2 M_{xy}}{\partial x \partial y} + \frac{\partial^2 M_y^n}{\partial y^2} + q_z = 0. \tag{2.54}$$

The last equation can be written in matrix notation as

$$\begin{bmatrix} \frac{\partial^2}{\partial x^2} & \frac{\partial^2}{\partial y^2} & \frac{2\partial^2}{\partial x \partial y} \end{bmatrix} \begin{bmatrix} M_x^n \\ M_y^n \\ M_{xy}^n \end{bmatrix} + q_z = 0, \tag{2.55}$$

**Table 2.3** Comparison of the equilibrium equations for thin beam (bending in the $xz$-plane) and thin plate elements

| Thin beam | Thin plate |
|---|---|
| Specific formulation | |
| $\dfrac{d^2 M_y(x)}{dx^2} + q_z(x) = 0$ | $\dfrac{\partial^2 M_x^n}{\partial x^2} + 2\dfrac{\partial^2 M_{xy}}{\partial x \partial y} + \dfrac{\partial^2 M_y^n}{\partial y^2} + q_z = 0$ |
| $\dfrac{dM_y(x)}{dx} = Q_z(x)$ | $\begin{bmatrix} \frac{\partial}{\partial x} & 0 & \frac{\partial}{\partial y} \\ 0 & \frac{\partial}{\partial y} & \frac{\partial}{\partial x} \end{bmatrix} \begin{bmatrix} M_x^n \\ M_y^n \\ M_{xy}^n \end{bmatrix} = \begin{bmatrix} Q_x^n \\ Q_y^n \end{bmatrix}$ |
| General formulation | |
| $\mathcal{L}_2^T M_y + q_z = 0$ | $\mathcal{L}_2^T M^n + q_z = 0$ |
| $\mathcal{L}_1^T M_y = Q_z$ | $\mathcal{L}_1^T M^n = Q^n$ |

or symbolically as

$$\mathcal{L}_2^T M^n + q_z = 0. \tag{2.56}$$

Furthermore, Eqs. (2.52) and (2.53) can be rearranged to obtain a relationship between the moments and shear forces:

$$\mathcal{L}_1^T M^n = Q^n, \tag{2.57}$$

where the first-order differential operator matrix $\mathcal{L}_1$ is given as:

$$\mathcal{L}_1 = \begin{bmatrix} \dfrac{\partial}{\partial x} & 0 \\ 0 & \dfrac{\partial}{\partial y} \\ \dfrac{\partial}{\partial y} & \dfrac{\partial}{\partial x} \end{bmatrix}. \tag{2.58}$$

Let us summarize at the end of this section the equilibrium equations for thin beams and thin plates, see Table 2.3. The similarity is particularly visible for the general formulation.

## 2.5 Differential Equations

Let us look first on the derivation of the differential equation for a thin beam. To this end, let us use the definition equation (2.24) for the internal bending moment and introduce Hooke's law (2.18) and the kinematics equation (2.4) to obtain the following expression:

$$M_y(x) = \int_A z\sigma_x \mathrm{d}A = \int_A zE\varepsilon_x \mathrm{d}A = \int_A zE\left(-z\frac{\mathrm{d}^2 u_z(x)}{\mathrm{d}x^2}\right)\mathrm{d}A. \qquad (2.59)$$

Rearranging the last equation gives

$$M_y(x) = -E\frac{\mathrm{d}^2 u_z(x)}{\mathrm{d}x^2}\underbrace{\int_A z^2 \mathrm{d}A}_{I_y}, \qquad (2.60)$$

or finally as the bending differential equation:

$$EI_y\frac{\mathrm{d}^2 u_z(x)}{\mathrm{d}x^2} = -M_y(x). \qquad (2.61)$$

It should be noted here that the integral in Eq. (2.60) is the so-called axial second moment of area or axial surface moment of 2nd order in the SI unit m$^4$. This factor is only dependent on the geometry of the cross section and is also a measure of the stiffness of a plane cross section against bending.

One-time differentiation of Eq. (2.61) with respect to the $x$-coordinate and considering of Eq. (2.32) gives the alternative formulation of the bending differential equation as:

$$EI_y\frac{\mathrm{d}^3 u_z(x)}{\mathrm{d}x^3} = -Q_z(x). \qquad (2.62)$$

Another one-time differentiation of the last expression and considering of Eq. (2.29) gives another alternative formulation of the bending differential equation as:

$$EI_y\frac{\mathrm{d}^4 u_z(x)}{\mathrm{d}x^4} = q_z(x). \qquad (2.63)$$

If we use again the formal operator symbol, i.e. $\mathcal{L}_2(\dots)$, to represent the second order derivative, i.e. $\frac{\mathrm{d}^2 \dots}{\mathrm{d}x^2}$, and use in addition the abbreviation $D = EI_y$ for the bending stiffness of the beam, Eq. (2.63) can be represented as:

$$\mathcal{L}_2^\mathrm{T}\left(D\mathcal{L}_2\left(u_z(x)\right)\right) - q_z(x) = 0. \qquad (2.64)$$

Let us look now on the derivation of the differential equation for a thin plate. To this end, let us combine the three equations for the resulting moments according to Eqs. (2.34)–(2.36) in matrix notation as

$$
M^n = \begin{bmatrix} M_x^n \\ M_y^n \\ M_{xy}^n \end{bmatrix} = \int_{-h/2}^{h/2} z \begin{bmatrix} \sigma_x \\ \sigma_y \\ \tau_{xy} \end{bmatrix} dz = \int_{-h/2}^{h/2} z\boldsymbol{\sigma}\, dz. \tag{2.65}
$$

Introducing Hooke's law (2.22) and the kinematics relation (2.17) gives for a constant elasticity matrix $C$

$$
M^n = -\int_{-h/2}^{h/2} z^2 C \mathcal{L}_2 u_z \, dz = -C\mathcal{L}_2 u_z \underbrace{\int_{-h/2}^{h/2} z^2 dz}_{\frac{h^3}{12}} = -\underbrace{\frac{h^3}{12} C}_{D_b} \mathcal{L}_2 u_z, \tag{2.66}
$$

where the plate elasticity matrix $D_b$ is given by

$$
D_b = \frac{h^3}{12} C = \underbrace{\frac{Eh^3}{12(1-v^2)}}_{D_b} \begin{bmatrix} 1 & v & 0 \\ v & 1 & 0 \\ 0 & 0 & \dfrac{1-v}{2} \end{bmatrix}, \tag{2.67}
$$

and $D_b = \frac{Eh^3}{12(1-v^2)}$ is the bending rigidity of the plate. Using the kinematics relation in the curvature form (see Eq. (2.17)), it can be stated that

$$
M^n = D_b \kappa. \tag{2.68}
$$

Introducing the moment-displacement relation (2.66) in the equilibrium equation (2.56) results in the plate bending differential equation in the form:

$$
\mathcal{L}_2^T (D_b \mathcal{L}_2 u_z) - q_z = 0. \tag{2.69}
$$

Using the definitions for $\mathcal{L}_2$ and $D_b$ given in Eqs. (2.55) and (2.67), the following classical form of the plate bending differential equation can be obtained:

$$
\frac{Eh^3}{12(1-v^2)} \left( \frac{\partial^4 u_z}{\partial x^4} + 2\frac{\partial^4 u_z}{\partial x^2 \partial y^2} + \frac{\partial^4 u_z}{\partial y^4} \right) = q_z. \tag{2.70}
$$

Let us summarize at the end of this section the differential equations for thin beams and thin plates, see Table 2.4. The similarity is particularly visible for the general formulation.

**Table 2.4** Comparison of the differential equations for thin beam (bending in the $xz$-plane) and thin plate elements

| Thin beam | Thin plate |
|---|---|
| Specific formulation | |
| $EI_y \dfrac{\mathrm{d}^4 u_z(x)}{\mathrm{d}x^4} - q_z(x) = 0$ | $\dfrac{Eh^3}{12(1-v^2)} \left( \dfrac{\partial^4 u_z}{\partial x^4} + 2 \dfrac{\partial^4 u_z}{\partial x^2 \partial y^2} + \dfrac{\partial^4 u_z}{\partial y^4} \right) - q_z = 0$ |
| General formulation | |
| $\mathcal{L}_2^{\mathrm{T}} \left( D\mathcal{L}_2 \left( u_z(x) \right) \right) - q_z(x) = 0$ | $\mathcal{L}_2^{\mathrm{T}} \left( \boldsymbol{D}_{\mathrm{b}} \mathcal{L}_2 u_z \right) - q_z = 0$ |

# References

1. Blaauwendraad J (2010) Plates and FEM: surprises and pitfalls. Springer, Dordrecht
2. Öchsner A (2021) Classical beam theories of structural mechanics. Springer, Cham
3. Reddy JN (2006) An introduction to the finite element method. McGraw Hill, Singapore
4. Timoshenko S, Woinowsky-Krieger S (1959) Theory of plates and shells. McGraw-Hill Book Company, New York
5. Ventsel E, Krauthammer T (2001) Thin plates and shells: theory, analysis, and applications. Marcel Dekker, New York
6. Wang CM, Reddy JN, Lee KH (2000) Shear deformable beams and plates: relationships with classical solutions. Elsevier, Oxford

# Chapter 3
# Thick Beams and Plates

**Abstract** This chapter presents the analytical description of thick beam and plate members. Based on the three basic equations of continuum mechanics, i.e., the kinematics relationship, the constitutive law and the equilibrium equation, the partial differential equations, which describe the physical problems, are derived. The presented theories for thick members consider, as one potential approach, a constant shear stress distribution over the thickness.

## 3.1 Introduction

The general configurations of a thick beam and a thick plate problem are shown in Fig. 3.1. The shear stress for beam bending problems has already been mentioned in Sect. 2.4. This shear stress is variable over the cross-section. As an example, one can mention the parabolic shear stress distribution over a rectangular cross section. Based on Hooke's law for a one-dimensional shear stress state, it can be derived that the shear strain has to exhibit a corresponding parabolic course. From the shear stress distribution in the cross-sectional area at location $x$ of the beam,[1] one calculates the acting shear force through integration as:

$$Q_z = \int_A \tau_{xz}(y, z)\, dA. \tag{3.1}$$

However, to simplify the problem, it is assumed for the so-called Timoshenko beam theory [9, 10] that an equivalent *constant* shear stress and strain act, see Fig. 3.2:

$$\tau_{xz}(y, z) \rightarrow \tau_{xz}. \tag{3.2}$$

---

[1] A closer analysis of the shear stress distribution in the cross-sectional area shows that the shear stress does not just alter over the height of the beam but also through the width of the beam. If the width of the beam is small when compared to the height, only a small change along the width occurs and one can assume in the first approximation a constant shear stress throughout the width: $\tau_{xz}(y, z) \rightarrow \tau_{xz}(z)$. See for example [2, 12].

A. Öchsner, *The Fundamental Equations of Beams and Plates*,
SpringerBriefs in Continuum Mechanics,
https://doi.org/10.1007/978-3-031-76276-5_3

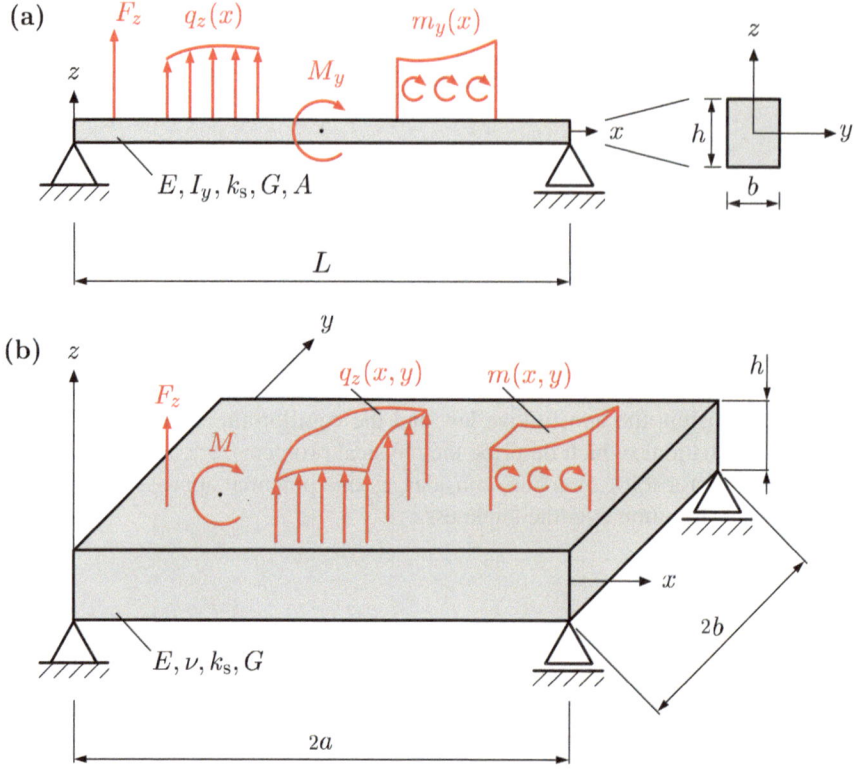

**Fig. 3.1** General configurations of **a** a thick beam (bending in the $xz$-plane) and **b** a thick plate problem

**Fig. 3.2** Shear stress distribution: **a** real distribution for a rectangular cross section and **b** Timoshenko's approximation

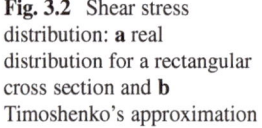

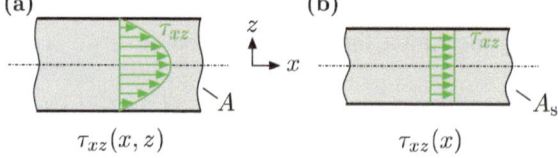

This constant shear stress results from the shear force, which acts in an equivalent cross-sectional area, the so-called shear area $A_s$:

$$\tau_{xz} = \frac{Q_z}{A_s}, \tag{3.3}$$

whereupon the relation between the shear area $A_s$ and the actual cross-sectional area $A$ is referred to as the shear correction factor $k_s$:

$$k_s = \frac{A_s}{A}. \tag{3.4}$$

**Table 3.1** Comparison of shear correction factor values for a rectangular cross section based on different approaches

| $k_s$ | Comment | References |
|---|---|---|
| $\frac{2}{3}$ | – | [9, 11] |
| $0.833 \left(= \frac{5}{6}\right)$ | $\nu = 0.0$ | [4] |
| 0.850 | $\nu = 0.3$ | |
| 0.870 | $\nu = 0.5$ | |

Different assumptions can be made to calculate the shear correction factor [4]. As an example, it can be demanded [1] that the elastic strain energy of the equivalent shear stress has to be identical with the energy, which results from the acting shear stress distribution in the actual cross-sectional-area. A comparison for a rectangular cross section is presented in Table 3.1.

Different geometric characteristics of simple geometric cross-sections—including the shear correction factor[2]—are collected in Table 3.2 [5, 15]. Further details regarding the shear correction factor for arbitrary cross-sections can be taken from [6].

It is obvious that the equivalent constant shear stress can alter along the center line of the beam, in case the shear force along the center line of the beam changes. The attribute 'constant' thus just refers to the cross-sectional area at location $x$ and the equivalent constant shear stress is therefore in general a function of the coordinate of length for the Timoshenko beam:

$$\tau_{xz} = \tau_{xz}(x). \tag{3.5}$$

The so-called Timoshenko beam can be generated by superposing a shear deformation on a Bernoulli beam according to Fig. 3.3.

## 3.2 Kinematics Equations

According to the derivation in Sect. 2.2, the kinematics relation can be derived for the beam with shear action, by considering the angle $\phi_y$ instead of the angle $\varphi_y$, see Figs. 2.2a and 3.4a.

Following an equivalent procedure as in Sect. 2.2, the corresponding relationships are obtained:

$$\sin \phi_y = \frac{u_x}{z} \approx \phi_y \quad \text{or} \quad u_x = +z\phi_y, \tag{3.6}$$

wherefrom, via the general relation for the strain, meaning $\varepsilon_x = du_x/dx$, the kinematics relation results through differentiation with respect to the $x$-coordinate:

---

[2] It should be noted that the so-called form factor for shear is also known in the literature. This results as the reciprocal of the shear correction factor.

**Table 3.2** Characteristics of different cross sections in the $yz$-plane. $I_y$ and $I_z$: axial second moments of area; $A$: cross-sectional area; $k_s$: shear correction factor. Adapted from [15]

| Cross-section | $I_y$ | $I_z$ | $A$ | $k_s$ |
|---|---|---|---|---|
| | $\dfrac{\pi R^4}{4}$ | $\dfrac{\pi R^4}{4}$ | $\pi R^2$ | $\dfrac{9}{10}$ |
| | $\pi R^3 t$ | $\pi R^3 t$ | $2\pi R t$ | 0.5 |
| | $\dfrac{bh^3}{12}$ | $\dfrac{hb^3}{12}$ | $hb$ | $\dfrac{5}{6}$ |
| | $\dfrac{h^2}{6}(ht_w + 3bt_f)$ | $\dfrac{b^2}{6}(bt_f + 3ht_w)$ | $2(bt_f + ht_w)$ | $\dfrac{2ht_w}{A}$ |
| | $\dfrac{h^2}{12}(ht_w + 6bt_f)$ | $\dfrac{b^3 t_f}{6}$ | $ht_w + 2bt_f$ | $\dfrac{ht_w}{A}$ |

$$\varepsilon_x = +z\frac{\mathrm{d}\phi_y}{\mathrm{d}x}. \tag{3.7}$$

Note that $\phi_y \rightarrow \varphi_y = -\frac{\mathrm{d}u_z}{\mathrm{d}x}$ results from neglecting the shear deformation and a relation according to Eq. (2.4) results as a special case. Furthermore, the following relation between the angles can be derived from Fig. 3.3c

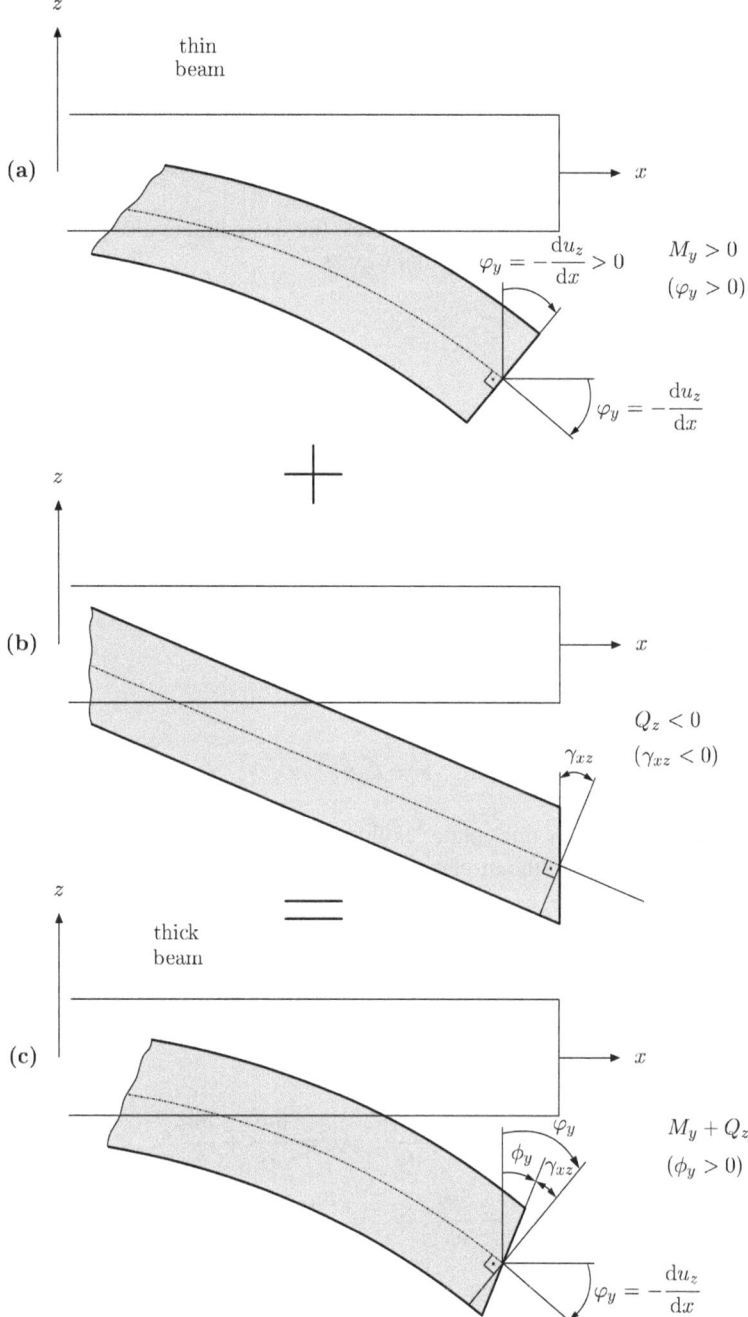

**Fig. 3.3** Superposition of the thin beam (**a**) and the shear deformation (**b**) to the thick (Timoshenko) beam (**c**) in the $xz$-plane. Note that the deformation is exaggerated for better illustration

$$\phi_y = \varphi_y + \gamma_{xz} = -\frac{\mathrm{d}u_z}{\mathrm{d}x} + \gamma_{xz}, \tag{3.8}$$

which complements the set of the kinematics relations. It needs to be remarked that at this point the so-called bending line was considered. Therefore, the displacement field $u_z$ is only a function of *one* variable: $u_z = u_z(x)$.

If we use the definitions of the curvatures, the basic kinematics equations (3.7) and (3.8) can be stated in a more formal way as

$$\kappa_x = \frac{\mathrm{d}\phi_y}{\mathrm{d}x}, \tag{3.9}$$

$$\kappa_{xz} = \frac{\mathrm{d}u_z}{\mathrm{d}x} + \phi_y, \tag{3.10}$$

or if we replace the common formulation of the first order derivative, i.e. $\frac{\mathrm{d}\dots}{\mathrm{d}x}$, by a formal operator symbol, i.e. $\mathcal{L}_1(\cdots)$, in matrix notation as:

$$\begin{bmatrix} \kappa_x \\ \kappa_{xz} \end{bmatrix} = \begin{bmatrix} 0 & \dfrac{\mathrm{d}}{\mathrm{d}x} \\ \dfrac{\mathrm{d}}{\mathrm{d}x} & 1 \end{bmatrix} \begin{bmatrix} u_z \\ \phi_y \end{bmatrix}, \tag{3.11}$$

or symbolically as

$$\boldsymbol{e} = \mathcal{L}_1 \boldsymbol{u}. \tag{3.12}$$

Let us consider now the thick plate.[3] Following the procedure outlined in Sect. 2.2, the relationships between the in-plane displacements and rotational angles are, see Fig. 3.4b:

$$u_x = +z\phi_y \; ; \quad u_y = -z\phi_x. \tag{3.13}$$

Expanding the classical relationships for a plane stress state as given in Eqs. (2.13)–(2.15) by two through-thickness shear strains, the following five relations can be given:

$$\varepsilon_x = \frac{\partial u_x}{\partial x} \; ; \; \varepsilon_y = \frac{\partial u_y}{\partial y} \; ; \; \gamma_{xy} = \frac{\partial u_x}{\partial y} + \frac{\partial u_y}{\partial x} \; ; \tag{3.14}$$

$$\gamma_{xz} = \frac{\partial u_x}{\partial z} + \frac{\partial u_z}{\partial x} \; ; \; \gamma_{yz} = \frac{\partial u_y}{\partial z} + \frac{\partial u_z}{\partial y}. \tag{3.15}$$

Considering the results from Eq. (3.13), the five kinematics relationships can be specialized to:

---

[3] The thick plate theory is also connected to the names of Reissner and Mindlin..

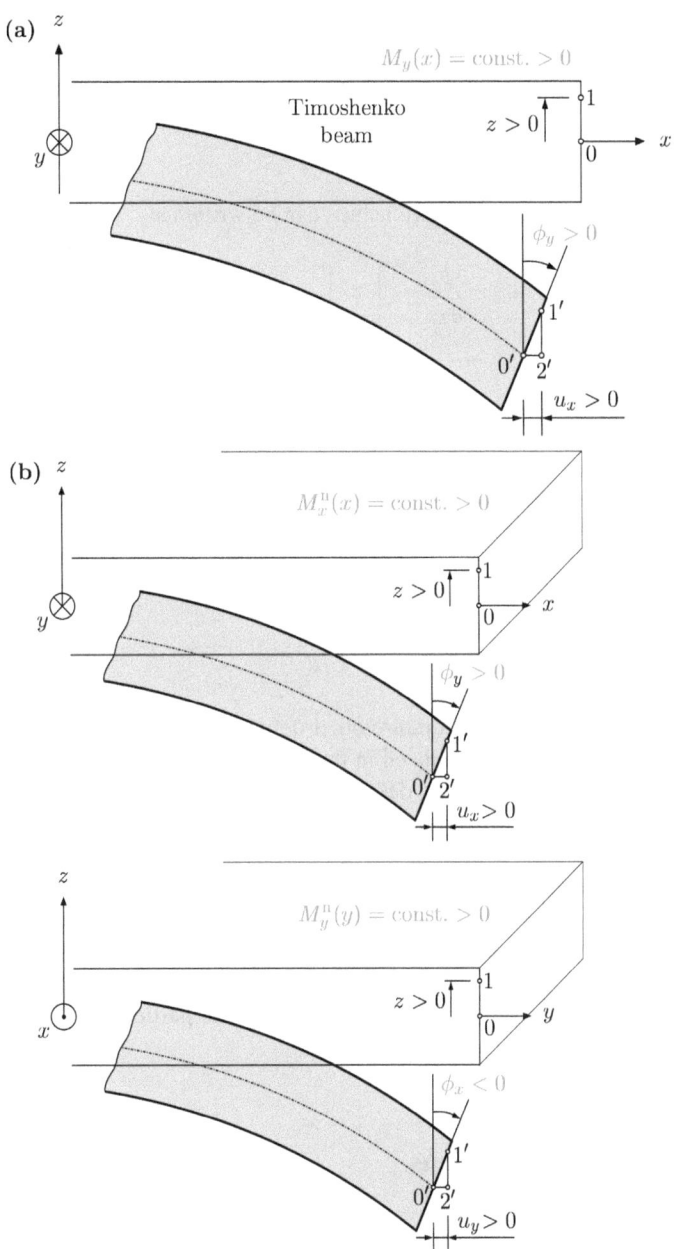

**Fig. 3.4** Configurations for the derivation of kinematics relations: **a** a thick beam (bending in the $xz$-plane) and **b** a thick plate problem

$$\varepsilon_x = z\frac{\partial \phi_y}{\partial x} = z\kappa_x \; ; \; \varepsilon_y = -z\frac{\partial \phi_x}{\partial y} = z\kappa_y \; ; \; \gamma_{xy} = z\left(\frac{\partial \phi_y}{\partial y} - \frac{\partial \phi_x}{\partial x}\right) = z\kappa_{xy} \; ;$$

$$\gamma_{xz} = \phi_y + \frac{\partial u_z}{\partial x} = \kappa_{xz} \; ; \; \gamma_{yz} = -\phi_x + \frac{\partial u_z}{\partial y} = \kappa_{yz}. \tag{3.16}$$

In matrix notation, these three relationships can be written as

$$\begin{bmatrix} \kappa_x \\ \kappa_y \\ \kappa_{xy} \\ \kappa_{xz} \\ \kappa_{yz} \end{bmatrix} = \begin{bmatrix} \dfrac{\partial \phi_y}{\partial x} \\ -\dfrac{\partial \phi_x}{\partial y} \\ \dfrac{\partial \phi_y}{\partial y} - \dfrac{\partial \phi_x}{\partial x} \\ \phi_y + \dfrac{\partial u_z}{\partial x} \\ -\phi_x + \dfrac{\partial u_z}{\partial y} \end{bmatrix} = \begin{bmatrix} 0 & 0 & \dfrac{\partial}{\partial x} \\ 0 & -\dfrac{\partial}{\partial y} & 0 \\ 0 & -\dfrac{\partial}{\partial x} & \dfrac{\partial}{\partial y} \\ \dfrac{\partial}{\partial x} & 0 & 1 \\ \dfrac{\partial}{\partial y} & -1 & 0 \end{bmatrix} \begin{bmatrix} u_z \\ \phi_x \\ \phi_y \end{bmatrix}, \tag{3.17}$$

or symbolically as

$$e = \mathcal{L}_1 u. \tag{3.18}$$

One may find in the scholarly literature other definitions of the rotational angles [3, 8, 13, 14]. The angle $\phi_y$ is introduced in the $xz$-plane, whereas $\phi_x$ is introduced in the $yz$-plane (see Fig. 3.4b). These definitions are closer to the classical definitions of the angles in the scope of finite elements but not conform with the definitions of the stress resultants (see $M_x^n$ and $M_y^n$ in Fig. 3.6b). Other definitions assume, for example, that the rotational angle $\varphi_x$ (now defined in the $xz$-plane) is positive if it leads to a positive displacement $u_x$ at the positive $z$-side of the neutral axis. The same definition holds for the angle $\varphi_y$ (now defined in the $yz$-plane).

Let us summarize at the end of this section the kinematics relations for thick beams and thick plates, see Table 3.3. The similarity is particularly visible for the general formulation.

## 3.3  Constitutive Equations

For the consideration of the constitutive relation of a thick beam, Hooke's law for a one-dimensional normal stress state and for a one-dimensional shear stress state (see Fig. 3.5a) is used:

$$\sigma_x = E\varepsilon_x, \tag{3.19}$$

$$\tau_{xz} = G\gamma_{xz}, \tag{3.20}$$

**Table 3.3** Comparison of the kinematics equations for thick beam (bending in the $xz$-plane) and thick plate elements

| Thick beam | Thick plate |
|---|---|
| Specific formulation | |
| $$\begin{bmatrix} \kappa_x \\ \kappa_{xz} \end{bmatrix} = \begin{bmatrix} 0 & \frac{d}{dx} \\ \frac{d}{dx} & 1 \end{bmatrix} \begin{bmatrix} u_z \\ \phi_y \end{bmatrix}$$ | $$\begin{bmatrix} \kappa_x \\ \kappa_y \\ \kappa_{xy} \\ \kappa_{xz} \\ \kappa_{yz} \end{bmatrix} = \begin{bmatrix} 0 & 0 & \frac{\partial}{\partial x} \\ 0 & -\frac{\partial}{\partial y} & 0 \\ 0 & -\frac{\partial}{\partial x} & \frac{\partial}{\partial y} \\ \frac{\partial}{\partial x} & 0 & 1 \\ \frac{\partial}{\partial y} & -1 & 0 \end{bmatrix} \begin{bmatrix} u_z \\ \phi_x \\ \phi_y \end{bmatrix}$$ |
| General formulation | |
| $e = \mathcal{L}_1 u$ | $e = \mathcal{L}_1 u$ |

**Fig. 3.5** Stresses acting on **a** a thick beam (bending in the $xz$-plane) and **b** a thick plate

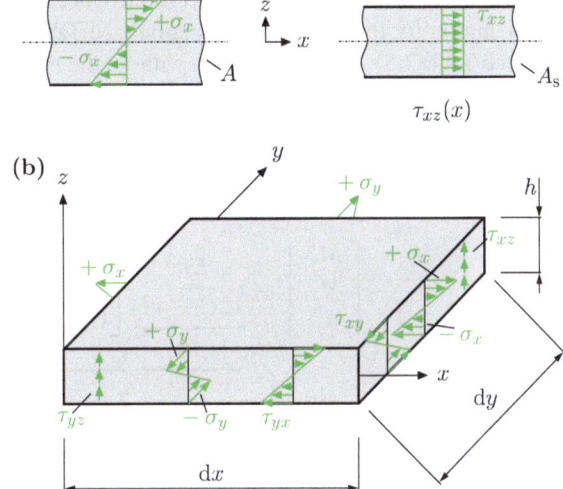

whereupon the shear modulus $G$ can be calculated for isotropic materials based on the Young's modulus $E$ and the Poisson's ratio $\nu$ as:

$$G = \frac{E}{2(1 + \nu)}. \tag{3.21}$$

The constitutive relations (3.19) and (3.20) can also be expressed in generalized stress-strain relationships, i.e. based on the international reactions and the corresponding curvatures:

$$M_y = E I_y \frac{d\phi_y}{dx} = E I_y \kappa_x, \tag{3.22}$$

$$Q_z = k_s A G \left( \frac{du_z}{dx} + \phi_y \right) = k_s A G \kappa_{xz}, \tag{3.23}$$

which can be expressed in matrix form as follows:

$$\begin{bmatrix} M_y \\ -Q_z \end{bmatrix} = \begin{bmatrix} E I_y & 0 \\ 0 & -k_s A G \end{bmatrix} \begin{bmatrix} \kappa_x \\ \kappa_{xz} \end{bmatrix}, \tag{3.24}$$

or symbolically as

$$s = De. \tag{3.25}$$

The minus sign in Eq. (3.24) has been introduced to guarantee the consistency in the $\mathcal{L}_1$-matrix for the kinematics equation (see Table 3.3) and the equilibrium equation (see Table 3.5). The formulation without the minus sign is treated separately in Sect. 3.6. This formulation without the minus sign results in two different differential operator matrices.

Let us start to assemble the constitutive equation for the thick plate[4] based on the plane stress formulation for a *thin* plate as given in Eq. (2.68) in terms of generalized quantities:

$$\begin{bmatrix} M_x^n \\ M_y^n \\ M_{xy}^n \end{bmatrix} = \underbrace{\frac{Eh^3}{12(1 - \nu^2)} \begin{bmatrix} 1 & \nu & 0 \\ \nu & 1 & 0 \\ 0 & 0 & \frac{1-\nu}{2} \end{bmatrix}}_{D_b} \begin{bmatrix} \kappa_x \\ \kappa_y \\ \kappa_{xy} \end{bmatrix}, \tag{3.26}$$

or under consideration of the definitions of the generalized strains $e$ as (see Eq. (3.17)):

$$\begin{bmatrix} M_x^n \\ M_y^n \\ M_{xy}^n \end{bmatrix} = \frac{Eh^3}{12(1 - \nu^2)} \begin{bmatrix} 1 & \nu & 0 \\ \nu & 1 & 0 \\ 0 & 0 & \frac{1-\nu}{2} \end{bmatrix} \begin{bmatrix} \dfrac{\partial \phi_y}{\partial x} \\ -\dfrac{\partial \phi_x}{\partial y} \\ \dfrac{\partial \phi_y}{\partial y} - \dfrac{\partial \phi_x}{\partial x} \end{bmatrix}. \tag{3.27}$$

In extension to the equations for the thick beam (see Eqs. (3.20) and (3.3)), the two through-thickness shear strains, which are approximated as constant values, can be related to the normalized shear forces by:

---

[4] We assume as in the case of the thick beam, i.e., the Timoshenko theory, that the shear stress (here: the two through-thickness stresses $\tau_{xz}$ and $\tau_{yz}$) is approximated by a constant value.

$$\begin{bmatrix} -Q_x^n \\ -Q_y^n \end{bmatrix} = -\underbrace{k_s Gh}_{D_s} \begin{bmatrix} \gamma_{xz} \\ \gamma_{yz} \end{bmatrix} = -\underbrace{k_s Gh \begin{bmatrix} 1 & 0 \\ 0 & 1 \end{bmatrix}}_{D_s} \begin{bmatrix} \gamma_{xz} \\ \gamma_{yz} \end{bmatrix},$$ (3.28)

where the minus sign was only introduced for formal reasons to have a certain consistency in the further derivations (see the consistency in the $\mathcal{L}_1$-matrix for the kinematics equation (see Table 3.3) and the equilibrium equation (see Table 3.5)).

Both equations for the constitutive contributions (see Eqs. (3.27) and (3.28)) can be combined to a single matrix form:

$$\begin{bmatrix} M_x^n \\ M_y^n \\ M_{xy}^n \\ -Q_x^n \\ -Q_y^n \end{bmatrix} = \left[ \begin{array}{c|c} \dfrac{Eh^3}{12(1-v^2)} \begin{bmatrix} 1 & v & 0 \\ v & 1 & 0 \\ 0 & 0 & \frac{1-v}{2} \end{bmatrix} & \begin{bmatrix} 0 & 0 \\ 0 & 0 \\ 0 & 0 \end{bmatrix} \\ \hline \begin{bmatrix} 0 & 0 & 0 \\ 0 & 0 & 0 \end{bmatrix} & -k_s Gh \begin{bmatrix} 1 & 0 \\ 0 & 1 \end{bmatrix} \end{array} \right] \begin{bmatrix} \dfrac{\partial \phi_y}{\partial x} \\ -\dfrac{\partial \phi_x}{\partial y} \\ \dfrac{\partial \phi_y}{\partial y} - \dfrac{\partial \phi_x}{\partial x} \\ \gamma_{xz} \\ \gamma_{yz} \end{bmatrix},$$ (3.29)

or as

$$\begin{bmatrix} M_x^n \\ M_y^n \\ M_{xy}^n \\ -Q_x^n \\ -Q_y^n \end{bmatrix} = \left[ \begin{array}{c|c} \dfrac{Eh^3}{12(1-v^2)} \begin{bmatrix} 1 & v & 0 \\ v & 1 & 0 \\ 0 & 0 & \frac{1-v}{2} \end{bmatrix} & \begin{bmatrix} 0 & 0 \\ 0 & 0 \\ 0 & 0 \end{bmatrix} \\ \hline \begin{bmatrix} 0 & 0 & 0 \\ 0 & 0 & 0 \end{bmatrix} & -k_s Gh \begin{bmatrix} 1 & 0 \\ 0 & 1 \end{bmatrix} \end{array} \right] \begin{bmatrix} \kappa_x \\ \kappa_y \\ \kappa_{xy} \\ \kappa_{xz} \\ \kappa_{yz} \end{bmatrix},$$ (3.30)

or symbolically as

$$s = De,$$ (3.31)

where $D$ is the plate elasticity matrix[5].

Let us summarize at the end of this section the constitutive equations for thick beams and thick plates, see Table 3.4. The similarity is particularly visible for the general formulation.

---

[5] This plate elasticity matrix should not be confused with the compliance matrix which is represented by the same symbol.

**Table 3.4** Comparison of the constitutive equations for thick beam (bending in the $xz$-plane) and thick plate elements

| Thick Beam | Thick Plate |
|---|---|
| Specific formulation | |

$$
\begin{bmatrix} M_y \\ -Q_z \end{bmatrix} = \begin{bmatrix} EI_y & 0 \\ 0 & -k_s AG \end{bmatrix} \begin{bmatrix} \kappa_x \\ \kappa_{xz} \end{bmatrix}
$$

$$
\begin{bmatrix} M_x^n \\ M_y^n \\ M_{xy}^n \\ -Q_x^n \\ -Q_y^n \end{bmatrix} = \begin{bmatrix} \dfrac{Eh^3}{12(1-\nu^2)} \begin{bmatrix} 1 & \nu & 0 \\ \nu & 1 & 0 \\ 0 & 0 & \frac{1-\nu}{2} \end{bmatrix} & \begin{bmatrix} 0 & 0 \\ 0 & 0 \\ 0 & 0 \end{bmatrix} \\ \begin{bmatrix} 0 & 0 & 0 \\ 0 & 0 & 0 \end{bmatrix} & -k_s Gh \begin{bmatrix} 1 & 0 \\ 0 & 1 \end{bmatrix} \end{bmatrix} \begin{bmatrix} \kappa_x \\ \kappa_y \\ \kappa_{xy} \\ \kappa_{xz} \\ \kappa_{yz} \end{bmatrix}
$$

| General formulation | |
|---|---|
| $s = De$ | $s = De$ |

## 3.4  Equilibrium Equations

The derivation of the equilibrium condition for the thick beam is identical with the derivation for the thin beam according to Sect. 2.4[6]:

$$
\frac{dQ_z(x)}{dx} = -q_z(x), \tag{3.32}
$$

$$
\frac{dM_y(x)}{dx} - Q_z(x) = -m_y(x). \tag{3.33}
$$

The last two equations can be combined in matrix from as

$$
\begin{bmatrix} 0 & \frac{d}{dx} \\ \frac{d}{dx} & 1 \end{bmatrix} \begin{bmatrix} M_y \\ -Q_z \end{bmatrix} + \begin{bmatrix} -q_z \\ +m_y \end{bmatrix} = \begin{bmatrix} 0 \\ 0 \end{bmatrix}, \tag{3.34}
$$

---

[6] We only assume in addition a distributed moment $m_y(x)$, which is rotating clockwise.

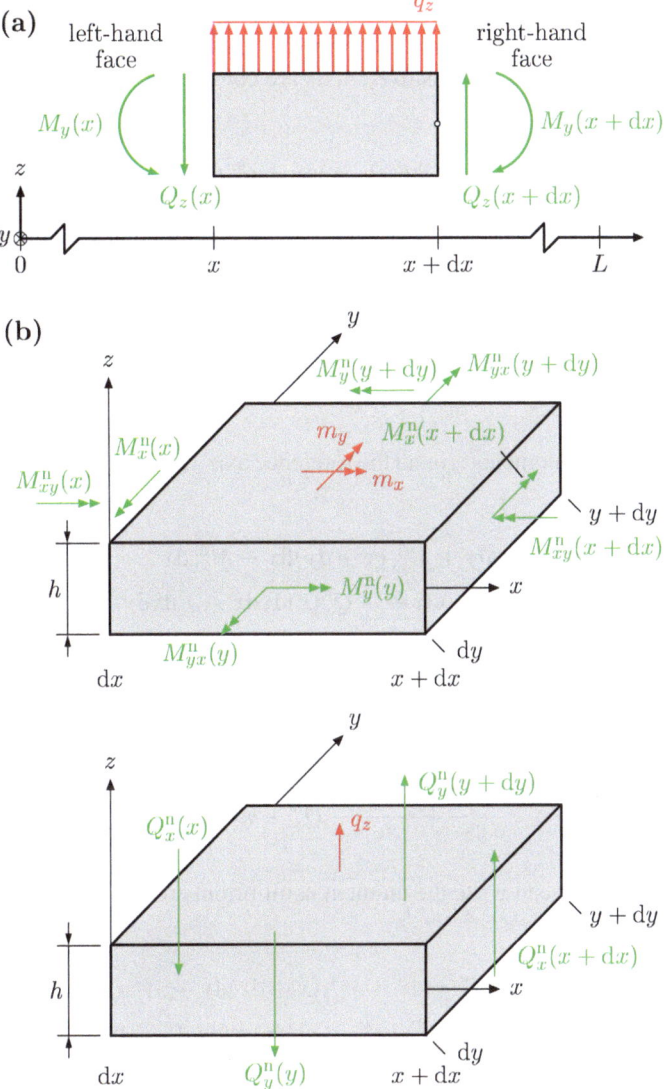

**Fig. 3.6** Infinitesimal elements with internal reactions: **a** a thick beam (bending in the $xz$-plane) and **b** a thick plate problem

or symbolically as

$$\mathcal{L}_1^{\mathrm{T}} s + b = 0. \qquad (3.35)$$

The derivation of the equilibrium equations for the thick plate follows the line of reasoning which was introduced in Sect. 2.4 for thin plates. In addition, we consider in the following the area distributed moments $m_x(x, y)$ and $m_y(x, y)$, see Fig. 3.6b.

The equilibrium condition will be determined in the following for the vertical forces. Assuming that the distributed force is constant $(q_z(x, y) \rightarrow q_z)$ and that forces in the direction of the positive $z$-axis are considered positive, the following results:

$$- Q_x^n(x)dy - Q_y^n(y)dx + Q_x^n(x + dx)dy + Q_y^n(y + dy)dx + q_z dx dy = 0.$$
(3.36)

Evaluating the shear forces at $x + dx$ and $y + dy$ in a Taylor's series of first order as outlined in Eqs. (2.40) and (2.41), the following expression for the vertical force equilibrium can be obtained:

$$\frac{\partial Q_x^n}{\partial x} + \frac{\partial Q_y^n}{\partial y} + q_z = 0.$$
(3.37)

The equilibrium of moments around the reference axis at $x + dx$ (positive if parallel to the $y$-axis) gives:

$$M_x^n(x + dx)dy - M_x^n(x)dy + M_{yx}^n(y + dy)dx - M_{yx}^n dx$$
$$- Q_y^n(y)dx \tfrac{dx}{2} + Q_y^n(y + dy)dx \tfrac{dx}{2} - Q_x^n(x)dy dx + q_z dx dy \tfrac{dx}{2} + m_y dx dy = 0.$$
(3.38)

Expanding the stress resultants at $x + dx$ and $y + dy$ into a Taylor's series of first order and neglecting the terms of third order gives finally:

$$\frac{\partial M_x^n}{\partial x} + \frac{\partial M_{xy}^n}{\partial y} - Q_x^n + m_y = 0.$$
(3.39)

In a similar way, we can write the moment equilibrium around the $x$-axis (with the reference axis at $y + dy$):

$$- M_y^n(y + dy)dx + M_y^n(y)dx - M_{xy}^n(x + dx)dy + M_{xy}^n(x)dy$$
$$- Q_x^n(x + dx)dy \tfrac{dy}{2} + Q_x^n(x)dy \tfrac{dy}{2} + Q_y^n(y)dx dy + m_x dx dy = 0.$$
(3.40)

Expanding the stress resultants at $x + dx$ and $y + dy$ into a Taylor's series of first order and neglecting the terms of third order gives finally:

$$\frac{\partial M_y^n}{\partial y} + \frac{\partial M_{xy}^n}{\partial x} - Q_y^n - m_x = 0.$$
(3.41)

The three equilibrium equations (see Eqs. (3.37), (3.39) and (3.41) can be written in matrix notation as

**Table 3.5** Comparison of the equilibrium equations for thick beam (bending in the $xz$-plane) and thick plate elements

| Thick beam | Thick plate |
|---|---|

Specific formulation

$$\begin{bmatrix} 0 & \frac{d}{dx} \\ \frac{d}{dx} & 1 \end{bmatrix} \begin{bmatrix} M_y \\ -Q_z \end{bmatrix} + \begin{bmatrix} -q_z \\ +m_y \end{bmatrix} = \begin{bmatrix} 0 \\ 0 \end{bmatrix}$$

$$\begin{bmatrix} 0 & 0 & 0 & \frac{\partial}{\partial x} & \frac{\partial}{\partial y} \\ 0 & -\frac{\partial}{\partial y} & -\frac{\partial}{\partial x} & 0 & -1 \\ \frac{\partial}{\partial x} & 0 & \frac{\partial}{\partial y} & 1 & 0 \end{bmatrix} \begin{bmatrix} M_x^n \\ M_y^n \\ M_{xy}^n \\ -Q_x^n \\ -Q_y^n \end{bmatrix} +$$

$$\begin{bmatrix} -q_z \\ m_x \\ m_y \end{bmatrix} = \begin{bmatrix} 0 \\ 0 \\ 0 \end{bmatrix}$$

General formulation

| $\mathcal{L}_1^{\mathrm{T}} s + b = 0$ | $\mathcal{L}_1^{\mathrm{T}} s + b = 0$ |
|---|---|

$$\begin{bmatrix} 0 & 0 & 0 & \frac{\partial}{\partial x} & \frac{\partial}{\partial y} \\ 0 & -\frac{\partial}{\partial y} & -\frac{\partial}{\partial x} & 0 & -1 \\ \frac{\partial}{\partial x} & 0 & \frac{\partial}{\partial y} & 1 & 0 \end{bmatrix} \begin{bmatrix} M_x^n \\ M_y^n \\ M_{xy}^n \\ -Q_x^n \\ -Q_y^n \end{bmatrix} + \begin{bmatrix} -q_z \\ m_x \\ m_y \end{bmatrix} = \begin{bmatrix} 0 \\ 0 \\ 0 \end{bmatrix}, \qquad (3.42)$$

or symbolically as

$$\mathcal{L}_1^{\mathrm{T}} s + b = 0. \qquad (3.43)$$

Let us summarize at the end of this section the equilibrium equations for thick beams and thick plates, see Table 3.5. The similarity is particularly visible for the general formulation.

## 3.5  Differential Equations

Introducing the constitutive equation (3.25) and the kinematics equation (3.12) in the equilibrium equation (3.35) gives the general rule for the derivation of the differential equation of the thick beam as:

$$\mathcal{L}_1^{\mathrm{T}} D \mathcal{L}_1 u + b = 0. \qquad (3.44)$$

The first matrix multiplication, i.e. $\mathcal{L}_1^{\mathrm{T}} D$, reads as[7]:

---

[7] Under the assumption of constant material and geometric properties.

$$
\begin{bmatrix} 0 & \dfrac{d}{dx} \\ \dfrac{d}{dx} & 1 \end{bmatrix} \begin{bmatrix} EI_y & 0 \\ 0 & -k_s AG \end{bmatrix} = \begin{bmatrix} 0 & -k_s AG \dfrac{d}{dx} \\ EI_y \dfrac{d}{dx} & -k_s AG \end{bmatrix}. \tag{3.45}
$$

The second matrix multiplication, i.e. $(\mathcal{L}_1^T D)\mathcal{L}_1$, reads

$$
\begin{bmatrix} 0 & -k_s AG \dfrac{d}{dx} \\ EI_y \dfrac{d}{dx} & -k_s AG \end{bmatrix} \begin{bmatrix} 0 & \dfrac{d}{dx} \\ \dfrac{d}{dx} & 1 \end{bmatrix} = \begin{bmatrix} -k_s AG \dfrac{d^2}{dx^2} & -k_s AG \dfrac{d}{dx} \\ -k_s AG \dfrac{d}{dx} & EI_y \dfrac{d^2}{dx^2} - k_s AG \end{bmatrix}, \tag{3.46}
$$

which finally results in the following matrix form of the differential equation:

$$
\begin{bmatrix} -k_s AG \dfrac{d^2}{dx^2} & -k_s AG \dfrac{d}{dx} \\ -k_s AG \dfrac{d}{dx} & EI_y \dfrac{d^2}{dx^2} - k_s AG \end{bmatrix} \begin{bmatrix} u_z \\ \phi_y \end{bmatrix} + \begin{bmatrix} -q_z \\ m_y \end{bmatrix} = \begin{bmatrix} 0 \\ 0 \end{bmatrix}, \tag{3.47}
$$

or symbolically as

$$
\mathcal{L}_1^T D \mathcal{L}_1 u + b = 0. \tag{3.48}
$$

Let us now look on the thick plate formulation. Introducing the constitutive equation (3.31) and the kinematics equation (3.18) in the equilibrium equation (3.43) gives the general rule for the derivation of the differential equation as:

$$
\mathcal{L}_1^T D \mathcal{L}_1 u + b = 0. \tag{3.49}
$$

The first matrix multiplication, i.e. $\mathcal{L}_1^T D$, reads as:

$$
\begin{bmatrix} 0 & 0 & 0 & \dfrac{\partial}{\partial x} & \dfrac{\partial}{\partial y} \\ 0 & -\dfrac{\partial}{\partial y} & -\dfrac{\partial}{\partial x} & 0 & -1 \\ \dfrac{\partial}{\partial x} & 0 & \dfrac{\partial}{\partial y} & 1 & 0 \end{bmatrix} \left[ \begin{array}{c|c} D_b \begin{bmatrix} 1 & v & 0 \\ v & 1 & 0 \\ 0 & 0 & \frac{1-v}{2} \end{bmatrix} & \begin{bmatrix} 0 & 0 \\ 0 & 0 \\ 0 & 0 \end{bmatrix} \\ \hline \begin{bmatrix} 0 & 0 & 0 \\ 0 & 0 & 0 \end{bmatrix} & -D_s \begin{bmatrix} 1 & 0 \\ 0 & 1 \end{bmatrix} \end{array} \right] =
$$

$$
\begin{bmatrix} 0 & 0 & 0 & -D_s \dfrac{\partial}{\partial x} & -D_s \dfrac{\partial}{\partial y} \\ -D_b \dfrac{\partial}{\partial y} v & -D_b \dfrac{\partial}{\partial y} & -D_b \dfrac{\partial}{\partial x} \dfrac{1-v}{2} & 0 & +D_s \\ D_b \dfrac{\partial}{\partial x} & D_b \dfrac{\partial}{\partial x} v & D_b \dfrac{\partial}{\partial y} \dfrac{1-v}{2} & -D_s & 0 \end{bmatrix}. \tag{3.50}
$$

The second matrix multiplication, i.e. $(\mathcal{L}_1^T D)\mathcal{L}_1$, reads

$$
\begin{bmatrix}
0 & 0 & 0 & -D_s\dfrac{\partial}{\partial x} & -D_s\dfrac{\partial}{\partial y} \\
-D_b\dfrac{\partial}{\partial y}v & -D_b\dfrac{\partial}{\partial y} & -D_b\dfrac{\partial}{\partial x}\dfrac{1-v}{2} & 0 & +D_s \\
D_b\dfrac{\partial}{\partial x} & D_b\dfrac{\partial}{\partial x}v & D_b\dfrac{\partial}{\partial y}\dfrac{1-v}{2} & -D_s & 0
\end{bmatrix}
\begin{bmatrix}
0 & 0 & \dfrac{\partial}{\partial x} \\
0 & -\dfrac{\partial}{\partial y} & 0 \\
0 & -\dfrac{\partial}{\partial x} & \dfrac{\partial}{\partial y} \\
\dfrac{\partial}{\partial x} & 0 & 1 \\
\dfrac{\partial}{\partial y} & -1 & 0
\end{bmatrix},
$$

$$(3.51)$$

which finally results in the following matrix form of the differential equation:

$$
\begin{bmatrix}
-D_s\left(\dfrac{\partial^2}{\partial x^2}+\dfrac{\partial^2}{\partial y^2}\right) & D_s\dfrac{\partial}{\partial y} & -D_s\dfrac{\partial}{\partial x} \\
D_s\dfrac{\partial}{\partial y} & D_b\left(\dfrac{1-v}{2}\dfrac{\partial^2}{\partial x^2}+\dfrac{\partial^2}{\partial y^2}\right)-D_s & -\dfrac{1+v}{2}D_b\dfrac{\partial^2}{\partial x\partial y} \\
-D_s\dfrac{\partial}{\partial x} & -\dfrac{1+v}{2}D_b\dfrac{\partial^2}{\partial x\partial y} & D_b\left(\dfrac{\partial^2}{\partial x^2}+\dfrac{1-v}{2}\dfrac{\partial^2}{\partial y^2}\right)-D_s
\end{bmatrix}
\times
$$

$$
\times
\begin{bmatrix} u_z \\ \phi_x \\ \phi_y \end{bmatrix}
+
\begin{bmatrix} -q_z \\ m_x \\ m_y \end{bmatrix}
=
\begin{bmatrix} 0 \\ 0 \\ 0 \end{bmatrix},
\qquad (3.52)
$$

or symbolically as

$$\mathcal{L}_1^T D\mathcal{L}_1 u + b = 0. \qquad (3.53)$$

Let us summarize at the end of this section the differential equations for thick beams and thick plates, see Table 3.6. The similarity is particularly visible for the general formulation.

## 3.6 Alternative Formulation

The general formulations of the basic equations for a thick beam as given in Tables 3.3, 3.4, 3.5 and 3.6 can be slightly modified to avoid the minus sign connected with $Q_z$. The kinematics equations (see Table 3.3) remain unchanged while the minus signs in the constitutive equation (see the matrix of generalized strains) can be eliminated:

**Table 3.6** Comparison of the differential equations for thick beam (bending in the $xz$-plane) and thick plate elements

| Thick beam | Thick plate |
|---|---|
| Specific formulation | |

Thick beam:

$$\left[\begin{array}{cc} -k_s AG \dfrac{\mathrm{d}^2}{\mathrm{d}x^2} & -k_s AG \dfrac{\mathrm{d}}{\mathrm{d}x} \\[2mm] -k_s AG \dfrac{\mathrm{d}}{\mathrm{d}x} & EI_y \dfrac{\mathrm{d}^2}{\mathrm{d}x^2} - k_s AG \end{array}\right]\left[\begin{array}{c} u_z \\ \phi_y \end{array}\right] + \left[\begin{array}{c} -q_z \\ m_y \end{array}\right] = \left[\begin{array}{c} 0 \\ 0 \end{array}\right]$$

Thick plate:

$$\left[\begin{array}{ccc} -D_s\left(\dfrac{\partial^2}{\partial x^2}+\dfrac{\partial^2}{\partial y^2}\right) & D_s\dfrac{\partial}{\partial y} & -D_s\dfrac{\partial}{\partial x} \\[3mm] D_s\dfrac{\partial}{\partial y} & D_b\left(\dfrac{1-\nu}{2}\dfrac{\partial^2}{\partial x^2}+\dfrac{\partial^2}{\partial y^2}\right)-D_s & \dfrac{1+\nu}{2}D_b\dfrac{\partial^2}{\partial x\partial y} \\[3mm] -D_s\dfrac{\partial}{\partial x} & \dfrac{1+\nu}{2}D_b\dfrac{\partial^2}{\partial x\partial y} & D_b\left(\dfrac{\partial^2}{\partial x^2}+\dfrac{1-\nu}{2}\dfrac{\partial^2}{\partial y^2}\right)-D_s \end{array}\right] \times$$

$$\times \left[\begin{array}{c} u_z \\ \phi_x \\ \phi_y \end{array}\right] + \left[\begin{array}{c} -q_z \\ m_x \\ m_y \end{array}\right] = \left[\begin{array}{c} 0 \\ 0 \\ 0 \end{array}\right]$$

| Thick beam | Thick plate |
|---|---|
| General Formulation | |
| $\mathcal{L}_1^{\mathrm{T}} D\mathcal{L}_1 u + b = 0$ | $\mathcal{L}_1^{\mathrm{T}} D\mathcal{L}_1 u + b = 0$ |

$$\begin{bmatrix} 1 & 0 \\ 0 & -1 \end{bmatrix} \begin{bmatrix} M_y \\ Q_z \end{bmatrix} = \begin{bmatrix} 1 & 0 \\ 0 & -1 \end{bmatrix} \begin{bmatrix} EI_y & 0 \\ 0 & k_s AG \end{bmatrix} \begin{bmatrix} \kappa_x \\ \kappa_{xz} \end{bmatrix}. \tag{3.54}$$

The diagonal matrix $\lceil 1 \; -1 \rfloor$ can be eliminated from the last equation to obtain the modified constitutive law in matrix notation:

$$\begin{bmatrix} M_y \\ Q_z \end{bmatrix} = \begin{bmatrix} EI_y & 0 \\ 0 & k_s AG \end{bmatrix} \begin{bmatrix} \kappa_x \\ \kappa_{xz} \end{bmatrix}. \tag{3.55}$$

The next step is to examine the equilibrium equation, i.e.,

$$\begin{bmatrix} 0 & \dfrac{d}{dx} \\ \dfrac{d}{dx} & 1 \end{bmatrix} \begin{bmatrix} M_y \\ -Q_z \end{bmatrix} + \begin{bmatrix} -q_z \\ +m_y \end{bmatrix} = \begin{bmatrix} 0 \\ 0 \end{bmatrix}, \tag{3.56}$$

or again re-written based on the diagonal matrices to extract the minus signs:

$$\begin{bmatrix} 0 & \dfrac{d}{dx} \\ \dfrac{d}{dx} & 1 \end{bmatrix} \begin{bmatrix} 1 & 0 \\ 0 & -1 \end{bmatrix} \begin{bmatrix} +M_y \\ +Q_z \end{bmatrix} + \begin{bmatrix} -1 & 0 \\ 0 & 1 \end{bmatrix} \begin{bmatrix} +q_z \\ +m_y \end{bmatrix} = \begin{bmatrix} 0 \\ 0 \end{bmatrix}. \tag{3.57}$$

Let us now multiply the first two matrices and then multiply the resulting equation with the $(2 \times 2)$ $\lceil -1 \; 1 \rfloor$ diagonal matrix from the left-hand side:

$$\begin{bmatrix} -1 & 0 \\ 0 & 1 \end{bmatrix} \underbrace{\left( \begin{bmatrix} 0 & \dfrac{d}{dx} \\ \dfrac{d}{dx} & 1 \end{bmatrix} \begin{bmatrix} 1 & 0 \\ 0 & -1 \end{bmatrix} \right)}_{\begin{bmatrix} 0 & -\dfrac{d}{dx} \\ \dfrac{d}{dx} & 1 \end{bmatrix}} \begin{bmatrix} M_y \\ Q_z \end{bmatrix} + \underbrace{\begin{bmatrix} -1 & 0 \\ 0 & 1 \end{bmatrix} \begin{bmatrix} -1 & 0 \\ 0 & 1 \end{bmatrix}}_{\begin{bmatrix} 1 & 0 \\ 0 & 1 \end{bmatrix}} \begin{bmatrix} q_z \\ m_y \end{bmatrix} = \begin{bmatrix} 0 \\ 0 \end{bmatrix}. \tag{3.58}$$

Or finally as the modified expression of the equilibrium equation:

$$\begin{bmatrix} 0 & \dfrac{d}{dx} \\ \dfrac{d}{dx} & -1 \end{bmatrix} \begin{bmatrix} M_y \\ Q_z \end{bmatrix} + \begin{bmatrix} q_z \\ m_y \end{bmatrix} = \begin{bmatrix} 0 \\ 0 \end{bmatrix}. \tag{3.59}$$

Combining the three basic equations, i.e., introducing the constitutive and the kinematics equation into the equilibrium, results again in the system of partial differential equations:

$$
\begin{bmatrix} 0 & \dfrac{\mathrm{d}}{\mathrm{d}x} \\ \dfrac{\mathrm{d}}{\mathrm{d}x} & -1 \end{bmatrix} \begin{bmatrix} EI_y & 0 \\ 0 & k_s AG \end{bmatrix} \begin{bmatrix} 0 & \dfrac{\mathrm{d}}{\mathrm{d}x} \\ \dfrac{\mathrm{d}}{\mathrm{d}x} & 1 \end{bmatrix} \begin{bmatrix} u_z \\ \phi_y \end{bmatrix} + \begin{bmatrix} q_z \\ m_y \end{bmatrix} = \begin{bmatrix} 0 \\ 0 \end{bmatrix},
\tag{3.60}
$$

or finally as:

$$
\begin{bmatrix} k_s AG \dfrac{\mathrm{d}^2}{\mathrm{d}x^2} & k_s AG \dfrac{\mathrm{d}}{\mathrm{d}x} \\ -k_s AG \dfrac{\mathrm{d}}{\mathrm{d}x} & EI_y \dfrac{\mathrm{d}^2}{\mathrm{d}x^2} - k_s AG \end{bmatrix} \begin{bmatrix} u_z \\ \phi_y \end{bmatrix} + \begin{bmatrix} q_z \\ m_y \end{bmatrix} = \begin{bmatrix} 0 \\ 0 \end{bmatrix}.
\tag{3.61}
$$

The modified basic equations, i.e., 'without the minus signs', are summarized for the thick beam in Table 3.7.

Following the same reasoning as in the case of the thick beam, the general formulations of the basic equations for a thick plate as given in Tables 3.3, 3.4, 3.5 and 3.6 can be slightly modified to avoid some esthetic appeals and to obtain more consistent representations, for example, of finite element matrices (cf. [7]). The kinematics equations remain unchanged while the minus signs in the constitutive equation (see the matrix of generalized strains) can be eliminated:

**Table 3.7**  Alternative formulations of the basic equations for a thick beam (bending in the $xz$-plane)

| Specific formulation | General formulation |
|---|---|
| Kinematics | |
| $\begin{bmatrix} \kappa_x \\ \kappa_{xz} \end{bmatrix} = \begin{bmatrix} 0 & \frac{\mathrm{d}}{\mathrm{d}x} \\ \frac{\mathrm{d}}{\mathrm{d}x} & 1 \end{bmatrix} \begin{bmatrix} u_z \\ \phi_y \end{bmatrix}$ | $e = \mathcal{L}_1 u$ |
| Constitution | |
| $\begin{bmatrix} M_y \\ Q_z \end{bmatrix} = \begin{bmatrix} EI_y & 0 \\ 0 & k_s AG \end{bmatrix} \begin{bmatrix} \kappa_x \\ \kappa_{xz} \end{bmatrix}$ | $s^* = D^* e$ |
| Equilibrium | |
| $\begin{bmatrix} 0 & \frac{\mathrm{d}}{\mathrm{d}x} \\ \frac{\mathrm{d}}{\mathrm{d}x} & -1 \end{bmatrix} \begin{bmatrix} M_y \\ Q_z \end{bmatrix} + \begin{bmatrix} q_z \\ m_y \end{bmatrix} = \begin{bmatrix} 0 \\ 0 \end{bmatrix}$ | $\mathcal{L}_{1*}^{\mathrm{T}} s^* + b^* = 0$ |
| PDE | |
| $\begin{bmatrix} k_s AG \frac{\mathrm{d}^2}{\mathrm{d}x^2} & k_s AG \frac{\mathrm{d}}{\mathrm{d}x} \\ -k_s AG \frac{\mathrm{d}}{\mathrm{d}x} & EI_y \frac{\mathrm{d}^2}{\mathrm{d}x^2} - k_s AG \end{bmatrix} \begin{bmatrix} u_z \\ \phi_y \end{bmatrix} + \begin{bmatrix} q_z \\ m_y \end{bmatrix} = \begin{bmatrix} 0 \\ 0 \end{bmatrix}$ | $\mathcal{L}_{1*}^{\mathrm{T}} D^* \mathcal{L}_1 u + b^* = 0$ |

$$
\begin{bmatrix} 1 & 0 & 0 & 0 & 0 \\ 0 & 1 & 0 & 0 & 0 \\ 0 & 0 & 1 & 0 & 0 \\ 0 & 0 & 0 & -1 & 0 \\ 0 & 0 & 0 & 0 & -1 \end{bmatrix} \begin{bmatrix} M_x^n \\ M_y^n \\ M_{xy}^n \\ Q_x^n \\ Q_y^n \end{bmatrix} = \begin{bmatrix} 1 & 0 & 0 & 0 & 0 \\ 0 & 1 & 0 & 0 & 0 \\ 0 & 0 & 1 & 0 & 0 \\ 0 & 0 & 0 & -1 & 0 \\ 0 & 0 & 0 & 0 & -1 \end{bmatrix} \left[ \begin{array}{c|c} D_b \left[ \ldots \right] & \left[ \ldots \right] \\ \hline \left[ \ldots \right] & D_s \left[ \ldots \right] \end{array} \right] \begin{bmatrix} \kappa_x \\ \kappa_y \\ \kappa_{xy} \\ \kappa_{xz} \\ \kappa_{yz} \end{bmatrix}.
$$

$$(3.62)$$

The diagonal matrix $\lceil 1\ 1\ 1\ -1\ -1 \rfloor$ can be eliminated from the last equation to obtain the modified constitutive law in matrix notation:

$$
\begin{bmatrix} M_x^n \\ M_y^n \\ M_{xy}^n \\ Q_x^n \\ Q_y^n \end{bmatrix} = \left[ \begin{array}{c|c} D_b \begin{bmatrix} 1 & v & 0 \\ v & 1 & 0 \\ 0 & 0 & \frac{1-v}{2} \end{bmatrix} & \begin{bmatrix} 0 & 0 \\ 0 & 0 \\ 0 & 0 \end{bmatrix} \\ \hline \begin{bmatrix} 0 & 0 & 0 \\ 0 & 0 & 0 \end{bmatrix} & D_s \begin{bmatrix} 1 & 0 \\ 0 & 1 \end{bmatrix} \end{array} \right] \begin{bmatrix} \kappa_x \\ \kappa_y \\ \kappa_{xy} \\ \kappa_{xz} \\ \kappa_{yz} \end{bmatrix}.
$$

$$(3.63)$$

The next step is to have a closer look on the equilibrium equation, i.e.,

$$
\begin{bmatrix} 0 & 0 & 0 & \dfrac{\partial}{\partial x} & \dfrac{\partial}{\partial y} \\ 0 & -\dfrac{\partial}{\partial y} & -\dfrac{\partial}{\partial x} & 0 & -1 \\ \dfrac{\partial}{\partial x} & 0 & \dfrac{\partial}{\partial y} & 1 & 0 \end{bmatrix} \begin{bmatrix} M_x^n \\ M_y^n \\ M_{xy}^n \\ -Q_x^n \\ -Q_y^n \end{bmatrix} + \begin{bmatrix} -q_z \\ m_x \\ m_y \end{bmatrix} = \begin{bmatrix} 0 \\ 0 \\ 0 \end{bmatrix},
$$

$$(3.64)$$

or again re-written based on the diagonal matrices to extract the minus signs:

$$
\begin{bmatrix} 0 & 0 & 0 & \dfrac{\partial}{\partial x} & \dfrac{\partial}{\partial y} \\ 0 & -\dfrac{\partial}{\partial y} & -\dfrac{\partial}{\partial x} & 0 & -1 \\ \dfrac{\partial}{\partial x} & 0 & \dfrac{\partial}{\partial y} & 1 & 0 \end{bmatrix} \begin{bmatrix} 1 & 0 & 0 & 0 & 0 \\ 0 & 1 & 0 & 0 & 0 \\ 0 & 0 & 1 & 0 & 0 \\ 0 & 0 & 0 & -1 & 0 \\ 0 & 0 & 0 & 0 & -1 \end{bmatrix} \begin{bmatrix} M_x^n \\ M_y^n \\ M_{xy}^n \\ Q_x^n \\ Q_y^n \end{bmatrix} + \begin{bmatrix} -1 & 0 & 0 \\ 0 & 1 & 0 \\ 0 & 0 & 1 \end{bmatrix} \begin{bmatrix} q_z \\ m_x \\ m_y \end{bmatrix} = \begin{bmatrix} 0 \\ 0 \\ 0 \end{bmatrix}.
$$

$$(3.65)$$

Let us now multiply the first two matrices and then multiply the resulting equation with the $(3 \times 3)$ diagonal matrix from the left-hand side:

$$
\begin{bmatrix} -1 & 0 & 0 \\ 0 & 1 & 0 \\ 0 & 0 & 1 \end{bmatrix}
\begin{bmatrix} 0 & 0 & 0 & -\dfrac{\partial}{\partial x} & -\dfrac{\partial}{\partial y} \\ 0 & -\dfrac{\partial}{\partial y} & -\dfrac{\partial}{\partial x} & 0 & 1 \\ \dfrac{\partial}{\partial x} & 0 & \dfrac{\partial}{\partial y} & -1 & 0 \end{bmatrix}
\begin{bmatrix} 1 & 0 & 0 & 0 & 0 \\ 0 & 1 & 0 & 0 & 0 \\ 0 & 0 & 1 & 0 & 0 \\ 0 & 0 & 0 & -1 & 0 \\ 0 & 0 & 0 & 0 & -1 \end{bmatrix}
\begin{bmatrix} M_x^{\mathrm{n}} \\ M_y^{\mathrm{n}} \\ M_{xy}^{\mathrm{n}} \\ Q_x^{\mathrm{n}} \\ Q_y^{\mathrm{n}} \end{bmatrix} +
$$

$$
\begin{bmatrix} -1 & 0 & 0 \\ 0 & 1 & 0 \\ 0 & 0 & 1 \end{bmatrix}
\begin{bmatrix} -1 & 0 & 0 \\ 0 & 1 & 0 \\ 0 & 0 & 1 \end{bmatrix}
\begin{bmatrix} q_z \\ m_x \\ m_y \end{bmatrix} =
\begin{bmatrix} 0 \\ 0 \\ 0 \end{bmatrix}. \tag{3.66}
$$

Or finally as the modified expression of the equilibrium equation:

$$
\begin{bmatrix} 0 & 0 & 0 & \dfrac{\partial}{\partial x} & \dfrac{\partial}{\partial y} \\ 0 & -\dfrac{\partial}{\partial y} & -\dfrac{\partial}{\partial x} & 0 & 1 \\ \dfrac{\partial}{\partial x} & 0 & \dfrac{\partial}{\partial y} & -1 & 0 \end{bmatrix}
\begin{bmatrix} M_x^{\mathrm{n}} \\ M_y^{\mathrm{n}} \\ M_{xy}^{\mathrm{n}} \\ Q_x^{\mathrm{n}} \\ Q_y^{\mathrm{n}} \end{bmatrix} +
\begin{bmatrix} q_z \\ m_x \\ m_y \end{bmatrix} =
\begin{bmatrix} 0 \\ 0 \\ 0 \end{bmatrix}. \tag{3.67}
$$

Combining the three basic equations results again in the system of partial differential equations:

$$
\begin{bmatrix}
D_{\mathrm{s}}\left(\dfrac{\partial^2}{\partial x^2} + \dfrac{\partial^2}{\partial y^2}\right) & -D_{\mathrm{s}}\dfrac{\partial}{\partial y} & D_{\mathrm{s}}\dfrac{\partial}{\partial x} \\[2ex]
D_{\mathrm{s}}\dfrac{\partial}{\partial y} & D_{\mathrm{b}}\left(\dfrac{1-\nu}{2}\dfrac{\partial^2}{\partial x^2} + \dfrac{\partial^2}{\partial y^2}\right) - D_{\mathrm{s}} & -\dfrac{1+\nu}{2}D_{\mathrm{b}}\dfrac{\partial^2}{\partial x \partial y} \\[2ex]
-D_{\mathrm{s}}\dfrac{\partial}{\partial x} & -\dfrac{1+\nu}{2}D_{\mathrm{b}}\dfrac{\partial^2}{\partial x \partial y} & D_{\mathrm{b}}\left(\dfrac{\partial^2}{\partial x^2} + \dfrac{1-\nu}{2}\dfrac{\partial^2}{\partial y^2}\right) - D_{\mathrm{s}}
\end{bmatrix} \times
$$

$$
\times \begin{bmatrix} u_z \\ \phi_x \\ \phi_y \end{bmatrix} +
\begin{bmatrix} q_z \\ m_x \\ m_y \end{bmatrix} =
\begin{bmatrix} 0 \\ 0 \\ 0 \end{bmatrix}. \tag{3.68}
$$

The modified basic equations, i.e., 'without the minus signs', are summarized for the thick plate in Table 3.8.

**Table 3.8** Alternative formulations of the basic equations for a thick plate

| Specific formulation | General formulation |
|---|---|
| **Kinematics** | |

$$\begin{bmatrix} \kappa_x \\ \kappa_y \\ \kappa_{xy} \\ \kappa_{xz} \\ \kappa_{yz} \end{bmatrix} = \begin{bmatrix} 0 & 0 & \frac{\partial}{\partial x} \\ 0 & -\frac{\partial}{\partial y} & 0 \\ 0 & -\frac{\partial}{\partial x} & \frac{\partial}{\partial y} \\ \frac{\partial}{\partial x} & 0 & 1 \\ \frac{\partial}{\partial y} & -1 & 0 \end{bmatrix} \begin{bmatrix} u_z \\ \phi_x \\ \phi_y \end{bmatrix} \qquad e = \mathcal{L}_1 u$$

| **Constitution** | |

$$\begin{bmatrix} M_x^{\mathrm{n}} \\ M_y^{\mathrm{n}} \\ M_{xy}^{\mathrm{n}} \\ Q_x^{\mathrm{n}} \\ Q_y^{\mathrm{n}} \end{bmatrix} = \begin{bmatrix} \underbrace{\frac{Eh^3}{12(1-\nu^2)}\begin{bmatrix} 1 & \nu & 0 \\ \nu & 1 & 0 \\ 0 & 0 & \frac{1-\nu}{2} \end{bmatrix}}_{D_{\mathrm{b}}} & \begin{bmatrix} 0 & 0 \\ 0 & 0 \\ 0 & 0 \end{bmatrix} \\ \begin{bmatrix} 0 & 0 & 0 \\ 0 & 0 & 0 \end{bmatrix} & \underbrace{k_{\mathrm{s}}Gh\begin{bmatrix} 1 & 0 \\ 0 & 1 \end{bmatrix}}_{D_{\mathrm{s}}} \end{bmatrix} \begin{bmatrix} \kappa_x \\ \kappa_y \\ \kappa_{xy} \\ \kappa_{xz} \\ \kappa_{yz} \end{bmatrix} \qquad s^* = D^* e$$

| **Equilibrium** | |

$$\begin{bmatrix} 0 & 0 & 0 & \frac{\partial}{\partial x} & \frac{\partial}{\partial y} \\ 0 & -\frac{\partial}{\partial y} & -\frac{\partial}{\partial x} & 0 & 1 \\ \frac{\partial}{\partial x} & 0 & \frac{\partial}{\partial y} & -1 & 0 \end{bmatrix} \begin{bmatrix} M_x^{\mathrm{n}} \\ M_y^{\mathrm{n}} \\ M_{xy}^{\mathrm{n}} \\ Q_x^{\mathrm{n}} \\ Q_y^{\mathrm{n}} \end{bmatrix} + \begin{bmatrix} q_z \\ m_x \\ m_y \end{bmatrix} = \begin{bmatrix} 0 \\ 0 \\ 0 \end{bmatrix} \qquad \mathcal{L}_{1*}^{\mathrm{T}} s^* + b^* = 0$$

| **PDE** | |

$$\begin{bmatrix} D_{\mathrm{s}}\left(\frac{\partial^2}{\partial x^2} + \frac{\partial^2}{\partial y^2}\right) & -D_{\mathrm{s}}\frac{\partial}{\partial y} & D_{\mathrm{s}}\frac{\partial}{\partial x} \\ D_{\mathrm{s}}\frac{\partial}{\partial y} & D_{\mathrm{b}}\left(\frac{1-\nu}{2}\frac{\partial^2}{\partial x^2} + \frac{\partial^2}{\partial y^2}\right) - D_{\mathrm{s}} & -\frac{1+\nu}{2}D_{\mathrm{b}}\frac{\partial^2}{\partial x\partial y} \\ -D_{\mathrm{s}}\frac{\partial}{\partial x} & -\frac{1+\nu}{2}D_{\mathrm{b}}\frac{\partial^2}{\partial x\partial y} & D_{\mathrm{b}}\left(\frac{\partial^2}{\partial x^2} + \frac{1-\nu}{2}\frac{\partial^2}{\partial y^2}\right) - D_{\mathrm{s}} \end{bmatrix} \begin{bmatrix} u_z \\ \phi_x \\ \phi_y \end{bmatrix} + \begin{bmatrix} q_z \\ m_x \\ m_y \end{bmatrix} = \begin{bmatrix} 0 \\ 0 \\ 0 \end{bmatrix} \qquad \mathcal{L}_{1*}^{\mathrm{T}} D^* \mathcal{L}_1 u + b^* = 0$$

# References

1. Bathe K-J (1996) Finite element procedures. Prentice-Hall, Upper Saddle River
2. Beer FP, Johnston ER Jr, DeWolf JT, Mazurek DF (2009) Mechanics of materials. McGraw-Hill, New York
3. Blaauwendraad J (2010) Plates and FEM: surprises and pitfalls. Springer, Dordrecht
4. Cowper GR (1966) The shear coefficient in Timoshenko's beam theory. J Appl Mech 33:335–340. https://doi.org/10.1115/1.3625046
5. Gere JM, Timoshenko SP (1991) Mechanics of materials. PWS-KENT Publishing Company, Boston
6. Gruttmann F, Wagner W (2001) Shear correction factors in Timoshenko's beam theory for arbitrary shaped cross-sections. Comput Mech 27:199–207. https://doi.org/10.1007/s004660100239
7. Öchsner A (2023) Computational statics and dynamics: an introduction based on the finite element method. Springer, Cham

8. Reddy JN (2006) An introduction to the finite element method. McGraw Hill, Singapore
9. Timoshenko SP (1921) On the correction for shear of the differential equation for transverse vibrations of prismatic bars. Philos Mag 41:744–746. https://doi.org/10.1080/14786442108636264
10. Timoshenko SP (1922) On the transverse vibrations of bars of uniform cross-section. Philos Mag 43:125–131. https://doi.org/10.1080/14786442208633855
11. Timoshenko S (1940) Strength of materials, part i: elementary theory and problems. D. Van Nostrand Company, New York
12. Timoshenko SP, Goodier JN (1970) Theory of easticity. McGraw-Hill, New York
13. Ventsel E, Krauthammer T (2001) Thin plates and shells: theory, analysis, and applications. Marcel Dekker, New York
14. Wang CM, Reddy JN, Lee KH (2000) Shear deformable beams and plates: relationships with classical solutions. Elsevier, Oxford
15. Weaver W Jr, Gere JM (1980) Matrix analysis of framed structures. Van Nostrand Reinhold Company, New York

# Chapter 4
# Comparison of the Approaches

**Abstract** This section provides a direct comparison between the thin and thick structural members under consideration of a formal description, this means based on differential operator symbols and generalized quantities.

Let us first compare the basic equations for a thin beam and a thin plate based on the differential operator symbols and generalized quantities [1]. Table 4.1 indicates that the same variables can be used to describe both structural members, however, the thin beam requires only the scalar version whereas the thin plate required a matrix version. Furthermore, it can be seen that the order of the differential operator, either as a scalar representation or as a matrix version, is of second order for both members. The comparison of the thin and a thick beam formulations presented in Table 4.2 indicates again that the same field variables (generalized stress and strain, as well as displacement) can be used at which the thin beam requires only a scalar representation and the thick beam is based on the matrix formulation of these quantities. In regards to the differential operator, the thin beam needs a scalar second order formulation whereas the thick beam needs a first order matrix formulation.

The comparison of the thick beam and thick plate formulations presented in Table 4.3 indicates exactly the same representations for both elements.

The comparison for the thin and thick plate in Table 4.4 shows that both set of equations require matrix formulations of the generalized stresses and strains. However, the thin plate requires only a scalar displacement whereas the thick plate needs a matrix version for the deformations. The same holds in regard to the representation of the external distributed loads. Both formulations require matrix versions of the differential operator, at which the thin plate requires a second order formulation and the thick plate is based on a first order operator matrix.

In conclusion it could be shown that all formulations reveal a very high similarity if generalized quantities and the differential operator symbol is used. For one case, i.e., the thick beam and thick plate, even a complete consistency is achieved. This formal description is also very useful for the derivation of numerical methods, e.g. the finite element method [2, 3].

© The Author(s), under exclusive license to Springer Nature Switzerland AG 2025
A. Öchsner, *The Fundamental Equations of Beams and Plates*,
SpringerBriefs in Continuum Mechanics,
https://doi.org/10.1007/978-3-031-76276-5_4

**Table 4.1**  Comparison of the basic equations for a thin beam and a thin plate (bending in $z$-direction)

| Thin beam | Thin plate |
|---|---|
| Kinematics | |
| $\varepsilon_x(x, z) = -z\mathcal{L}_2\left(u_z(x)\right)$ | $\boldsymbol{\varepsilon}(x, y, z) = -z\boldsymbol{\mathcal{L}}_2 u_z(x, y)$ |
| $\kappa(x) = -\mathcal{L}_2 u_z(x)$ | $\boldsymbol{\kappa}(x, y) = -\boldsymbol{\mathcal{L}}_2 u_z(x, y)$ |
| Constitution | |
| $\sigma_x(x, z) = C\varepsilon_x(x, z)$ | $\boldsymbol{\sigma}(x, y, z) = \boldsymbol{C}\boldsymbol{\varepsilon}(x, y, z)$ |
| $M_y(x) = D\kappa(x)$ | $\boldsymbol{M}^{\mathrm{n}}(x, y) = \boldsymbol{D}_{\mathrm{b}}\boldsymbol{\kappa}(x, y)$ |
| Equilibrium | |
| $\mathcal{L}_2^{\mathrm{T}}\left(M_y(x)\right) + q_z(x) = 0$ | $\boldsymbol{\mathcal{L}}_2^{\mathrm{T}}\boldsymbol{M}^{\mathrm{n}}(x, y) + q_z(x, y) = 0$ |
| PDE | |
| $\mathcal{L}_2^{\mathrm{T}}\left(D\mathcal{L}_2\left(u_z(x)\right)\right) - q_z(x) = 0$ | $\boldsymbol{\mathcal{L}}_2^{\mathrm{T}}\left(\boldsymbol{D}_{\mathrm{b}}\boldsymbol{\mathcal{L}}_2 u_z(x, y)\right) - q_z(x, y) = 0$ |

**Table 4.2**  Comparison of the basic equations for a thin and a thick beam (bending in $z$-direction)

| Thin beam | Thick beam |
|---|---|
| Kinematics | |
| $\varepsilon_x(x, z) = -z\mathcal{L}_2\left(u_z(x)\right)$ | |
| $\underbrace{\kappa(x) = -\mathcal{L}_2 u_z(x)}_{e}$ | $\mathbf{e} = \boldsymbol{\mathcal{L}}_1\mathbf{u}$ |
| Constitution | |
| $\sigma_x(x, z) = C\varepsilon_x(x, z)$ | |
| $\underbrace{M_y(x)}_{s} = D\,\underbrace{\kappa(x)}_{e}$ | $\mathbf{s} = \mathbf{D}\mathbf{e}$ |
| Equilibrium | |
| $\underbrace{\mathcal{L}_2^{\mathrm{T}}\left(M_y(x)\right)}_{s} + \underbrace{q_z(x)}_{b} = 0$ | $\boldsymbol{\mathcal{L}}_1^{\mathrm{T}}\mathbf{s} + \mathbf{b} = 0$ |
| PDE | |
| $\mathcal{L}_2^{\mathrm{T}}\left(D\mathcal{L}_2\left(u_z(x)\right)\right) - \underbrace{q_z(x)}_{b} = 0$ | $\boldsymbol{\mathcal{L}}_1^{\mathrm{T}}\mathbf{D}\boldsymbol{\mathcal{L}}_1\mathbf{u} + \mathbf{b} = 0$ |

**Table 4.3** Comparison of the basic equations for a thick beam and a thick plate (bending in $z$-direction)

| Thick beam | Thick plate |
|---|---|
| Kinematics | |
| $e = \mathcal{L}_1 u$ | $e = \mathcal{L}_1 u$ |
| Constitution | |
| $s = De$ | $s = De$ |
| Equilibrium | |
| $\mathcal{L}_1^{\mathrm{T}} s + b = 0$ | $\mathcal{L}_1^{\mathrm{T}} s + b = 0$ |
| PDE | |
| $\mathcal{L}_1^{\mathrm{T}} D \mathcal{L}_1 u + b = 0$ | $\mathcal{L}_1^{\mathrm{T}} D \mathcal{L}_1 u + b = 0$ |

**Table 4.4** Comparison of the basic equations for a thin and a thick plate (bending in $z$-direction)

| Thin Plate | Thick plate |
|---|---|
| Kinematics | |
| $\underbrace{\kappa(x, y) = -\mathcal{L}_2 u_z(x, y)}_{e}$ | $e = \mathcal{L}_1 u$ |
| Constitution | |
| $\underbrace{M^{\mathrm{n}}(x, y)}_{s} = D_{\mathrm{b}} \underbrace{\kappa(x, y)}_{e}$ | $s = De$ |
| Equilibrium | |
| $\mathcal{L}_2^{\mathrm{T}} \underbrace{M^{\mathrm{n}}(x, y)}_{s} + \underbrace{q_z(x, y)}_{b} = 0$ | $\mathcal{L}_1^{\mathrm{T}} s + b = 0$ |
| PDE | |
| $\mathcal{L}_2^{\mathrm{T}} (D_{\mathrm{b}} \mathcal{L}_2 u_z(x, y)) - \underbrace{q_z(x, y)}_{b} = 0$ | $\mathcal{L}_1^{\mathrm{T}} D \mathcal{L}_1 u + b = 0$ |

# References

1. Altenbach H, Öchsner A (eds) (2020) Encyclopedia of continuum mechanics. Springer, Berlin
2. Öchsner A (2023) Computational statics and dynamics: an introduction based on the finite element method. Springer, Cham
3. Zienkiewicz OC, Taylor RL (2000) The finite element method. In: Solid Mechanics, vol 2. Butterworth-Heinemann, Oxford

# Appendix
# Bending of Beams in the $xy$-Plane

Let us collect in this section the general equations for beams, which are only bending in the $xy$-plane. The detailed derivation of the fundamental equations is not presented here, however, a simple adjustment of the course of reasoning presented in Chaps. 2 and 3 can be done or the corresponding literature consulted, e.g. [1].

## A.1 Thin Beams

The general configuration for the bending of a thin beam in the $xy$-plane is presented in Fig. A.1. It should be noted here that a positive rotation of a moment is now counterclockwise and that forces are acting in $y$-direction.

The elementary basic equations for the bending of a beam in the $xy$-plane for an arbitrary shear force $Q_y(x)$ and moment loading $M_z(x)$ are summarized in Table A.1.

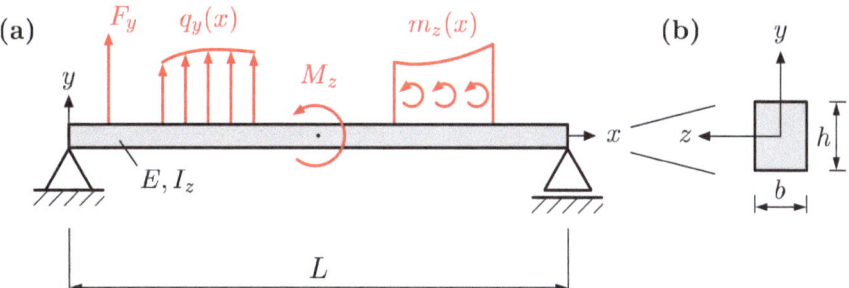

**Fig. A.1** General configuration for thin beam problems in the $xy$-plane: **a** example of boundary conditions and external loads (drawn in their positive directions); **b** cross-sectional area

© The Editor(s) (if applicable) and The Author(s), under exclusive license to Springer
Nature Switzerland AG 2025
A. Öchsner, *The Fundamental Equations of Beams and Plates*,
SpringerBriefs in Continuum Mechanics,
https://doi.org/10.1007/978-3-031-76276-5

## A.2 Thick Beams

The general configuration for the bending of a thick beam in the $xy$-plane is presented in Fig. A.2. It should be noted here that a positive rotation of a moment is now counterclockwise and that forces are acting in $y$-direction.

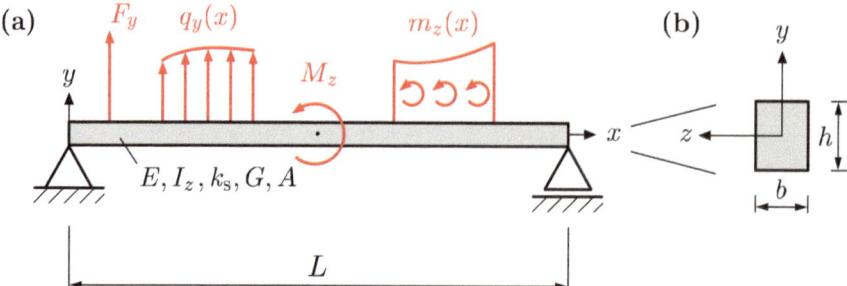

**Fig. A.2** General configuration for thick beam problems in the $xy$-plane: **a** example of boundary conditions and external loads (drawn in their positive directions); **b** cross-sectional area

**Table A.1** Elementary basic equations for the bending of a thin beam in the $xy$-plane. The differential equations are given under the assumption of constant bending stiffness $EI_z$

| Specific formulation | General formulation |
|---|---|
| Kinematics | |
| $\varepsilon_x(x, y) = -y\dfrac{\mathrm{d}^2 u_y(x)}{\mathrm{d}x^2}$ | $\varepsilon_x(x, y) = -y\mathcal{L}_2\left(u_y(x)\right) = -y\kappa_x(x)$ |
| Constitution | |
| $\sigma_x(x, y) = E\varepsilon_x(x, y)$ | $\sigma_x(x, y) = C\varepsilon_x(x, y)$ |
| $M_z(x) = EI_z\kappa_x(x)$ | $M_z(x) = D\kappa_x(x)$ |
| Equilibrium | |
| $\dfrac{\mathrm{d}Q_y(x)}{\mathrm{d}x} + q_y(x) = 0$ | |
| $\dfrac{\mathrm{d}^2 M_z(x)}{\mathrm{d}x^2} - q_y(x) + \dfrac{\mathrm{d}m_z(x)}{\mathrm{d}x} = 0$ | $\mathcal{L}_2^{\mathrm{T}} M_z(x) - q_y(x) + \dfrac{\mathrm{d}m_z(x)}{\mathrm{d}x} = 0$ |
| Stress | |
| $\sigma_x(x, y) = -\dfrac{M_z(x)}{I_z}y(x)$ | |
| PDE | |
| $EI_z\dfrac{\mathrm{d}^2 u_y(x)}{\mathrm{d}x^2} = M_z(x)$ | |
| $EI_z\dfrac{\mathrm{d}^3 u_y(x)}{\mathrm{d}x^3} = -Q_y(x) - m_z(x)$ | |
| $EI_z\dfrac{\mathrm{d}^4 u_y(x)}{\mathrm{d}x^4} = q_y(x) - \dfrac{\mathrm{d}m_z(x)}{\mathrm{d}x}$ | $\mathcal{L}_2^{\mathrm{T}} D\mathcal{L}_2 u_y(x) - q_y(x) + \dfrac{\mathrm{d}m_z(x)}{\mathrm{d}x} = 0$ |

The elementary basic equations for the bending of a beam in the $xy$-plane for an arbitrary shear force $Q_y(x)$ and moment loading $M_z(x)$ are summarized in Table A.2.

**Table A.2** Different formulations of the basic equations for a Timoshenko beam (bending in the $xy$-plane; $x$-axis along the principal beam axis). $E$: Young's modulus; $G$: shear modulus; $A$: cross-sectional area; $I_z$: second moment of area; $k_s$: shear correction factor; $q_y$: length-specific distributed force; $m_z$: length-specific distributed moment; $e$: generalized strains; $s$: generalized stresses

| Specific formulation | General formulation |
|---|---|
| **Kinematics** | |
| $\begin{bmatrix} \kappa_x \\ \kappa_{xy} \end{bmatrix} = \begin{bmatrix} 0 & \frac{\mathrm{d}}{\mathrm{d}x} \\ \frac{\mathrm{d}}{\mathrm{d}x} & -1 \end{bmatrix} \begin{bmatrix} u_y \\ \phi_z \end{bmatrix}$ | $e = \mathcal{L}_1 u$ |
| **Constitution** | |
| $\begin{bmatrix} M_z \\ -Q_y \end{bmatrix} = \begin{bmatrix} E I_z & 0 \\ 0 & -k_s A G \end{bmatrix} \begin{bmatrix} \kappa_x \\ \kappa_{xy} \end{bmatrix}$ | $s = De$ |
| **Equilibrium** | |
| $\begin{bmatrix} 0 & \frac{\mathrm{d}}{\mathrm{d}x} \\ \frac{\mathrm{d}}{\mathrm{d}x} & -1 \end{bmatrix} \begin{bmatrix} M_z \\ -Q_y \end{bmatrix} + \begin{bmatrix} q_y \\ m_z \end{bmatrix} = \begin{bmatrix} 0 \\ 0 \end{bmatrix}$ | $\mathcal{L}_1^{\mathrm{T}} s + b = 0$ |
| **PDEs** | |
| $\begin{bmatrix} \frac{\mathrm{d}}{\mathrm{d}x}\left[ k_s G A \left( \frac{\mathrm{d}u_y}{\mathrm{d}x} - \phi_z \right) \right] \\ \frac{\mathrm{d}}{\mathrm{d}x}\left( E I_z \frac{\mathrm{d}\phi_z}{\mathrm{d}x} \right) + k_s G A \left( \frac{\mathrm{d}u_y}{\mathrm{d}x} - \phi_z \right) \end{bmatrix} + \begin{bmatrix} q_y \\ m_z \end{bmatrix} = \begin{bmatrix} 0 \\ 0 \end{bmatrix}$ | $\mathcal{L}_1^{\mathrm{T}} D \mathcal{L}_1 u + b = 0$ |

**Table A.3** Alternative formulations of the basic equations for a Timoshenko beam (bending in the $xy$-plane; $x$-axis along the principal beam axis). $E$: Young's modulus; $G$: shear modulus; $A$: cross-sectional area; $I_z$: second moment of area; $k_s$: shear correction factor; $q_y$: length-specific distributed force; $m_z$: length-specific distributed moment; $e$: generalized strains; $s^*$: generalized stresses

| Specific formulation | General formulation |
|---|---|
| **Kinematics** | |
| $\begin{bmatrix} \kappa_x \\ \kappa_{xy} \end{bmatrix} = \begin{bmatrix} 0 & \frac{\mathrm{d}}{\mathrm{d}x} \\ \frac{\mathrm{d}}{\mathrm{d}x} & -1 \end{bmatrix} \begin{bmatrix} u_y \\ \phi_z \end{bmatrix}$ | $e = \mathcal{L}_1 u$ |
| **Constitution** | |
| $\begin{bmatrix} M_z \\ Q_y \end{bmatrix} = \begin{bmatrix} E I_z & 0 \\ 0 & k_s A G \end{bmatrix} \begin{bmatrix} \kappa_x \\ \kappa_{xy} \end{bmatrix}$ | $s^* = D^* e$ |
| **Equilibrium** | |
| $\begin{bmatrix} 0 & \frac{\mathrm{d}}{\mathrm{d}x} \\ \frac{\mathrm{d}}{\mathrm{d}x} & 1 \end{bmatrix} \begin{bmatrix} M_z \\ Q_y \end{bmatrix} + \begin{bmatrix} q_y \\ m_z \end{bmatrix} = \begin{bmatrix} 0 \\ 0 \end{bmatrix}$ | $\mathcal{L}_{1*}^{\mathrm{T}} s^* + b = 0$ |
| **PDE** | |
| $\begin{bmatrix} \frac{\mathrm{d}}{\mathrm{d}x}\left[ k_s G A \left( \frac{\mathrm{d}u_y}{\mathrm{d}x} - \phi_z \right) \right] \\ \frac{\mathrm{d}}{\mathrm{d}x}\left( E I_z \frac{\mathrm{d}\phi_z}{\mathrm{d}x} \right) + k_s G A \left( \frac{\mathrm{d}u_y}{\mathrm{d}x} - \phi_z \right) \end{bmatrix} + \begin{bmatrix} q_y \\ m_z \end{bmatrix} = \begin{bmatrix} 0 \\ 0 \end{bmatrix}$ | $\mathcal{L}_{1*}^{\mathrm{T}} D^* \mathcal{L}_1 u + b = 0$ |

We can see from this table that the matrix for the differential operators is identical for the kinematics and the equilibrium equation. However, we have a minus sign connected with the internal shear force (see the equation for the constitution and equilibrium equation).

To avoid the minus sign connected with the internal shear force, an alternative set of basic equations can be provided, see Table A.3. However, this alternative formulation results in slightly different matrices for the differential operators for the kinematics and the equilibrium equation.

# References

1. Altenbach H, Öchsner A (eds) (2020) Encyclopedia of continuum mechanics. Springer, Berlin

# Index

**A**
Axial second moment of area, 22

**B**
Bending
  pure, 7
  $xy$-plane, 53
  $xz$-plane, 7, 25
Bending rigidity of a plate, 23, 34
Bending stiffness of a beam, 22

**C**
Compliance matrix, 13
Constitutive equation, 3
  thick beam, 32
  thick plate, 35
  thin beam, 12
  thin plate, 12
Continuum mechanical modeling, 3
Coordinate system, 4

**D**
Differential equation, 3
  thick beam, 39
  thick plate, 40
  thin beam, 22
  thin plate, 23

**E**
Elasticity matrix, 13
Equilibrium equation, 3
  thick beam, 36
  thick plate, 38
  thin beam, 14
  thin plate, 20
Euler-Bernoulli beam, *see* Thin beam
External loads
  thick beam, 26
  thick plate, 26
  thin beam, 5, 8
  thin plate, 5, 8

**G**
Generalized strain, 33, 34, 41, 44, 56
Generalized stress, 33, 56

**I**
Internal reactions
  thick beam, 37
  thick plate, 37
  thin beam, 15, 18
  thin plate, 15, 18

**K**
Kinematics relation, 3
  thick beam, 27
  thick plate, 30

© The Editor(s) (if applicable) and The Author(s), under exclusive license to Springer
Nature Switzerland AG 2025
A. Öchsner, *The Fundamental Equations of Beams and Plates*,
SpringerBriefs in Continuum Mechanics,
https://doi.org/10.1007/978-3-031-76276-5

57

thin beam, 10
thin plate, 11
Kirchhoff plate theory, *see* Thin plate

**M**
Modulus of elasticity, 12

**P**
Poisson's ratio, 33

**R**
Reissner-Mindlin plate theory, *see* Thick plate

**S**
Shear area, 26
Shear correction factor, 26, 28
Shear-flexible models, 6
Shear force, 25
Shear modulus, 33
Shear stress
  equivalent, 26
Shear-rigid models, 6
Stress distributions
  thick beam, 26, 33
  thick plate, 33
  thin beam, 13, 26
  thin plate, 13

Stress resultants, 14, 17

**T**
Thick beam
  constitutive equation, 32
  constitutive relation, 32
  differential equation, 39
  equilibrium condition, 36
  kinematics relation, 27
Thick plate
  constitutive equation, 35
  differential equation, 40
  equilibrium equation, 38
  kinematics relation, 30
Thin beam
  constitutive equation, 12
  differential equation, 22
  equilibrium equation, 14
  kinematics relation, 10
Thin member
  definition, 7
Thin plate
  constitutive equation, 12
  differential equation, 23
  equilibrium equation, 20
  kinematics relation, 11
Timoshenko beam, *see* Thick beam

**Y**
Young's modulus, *see* Modulus of elasticity